Lecture Notes in Mathematics

Edited by A. Dold and B. Eckmann

Subseries: Mathematisches Institut der Universität und Max-Planck-Institut für Mathematik, Bonn – vol. 13

Adviser: F. Hirzebruch

1301

Norbert Schappacher

Periods of Hecke Characters

Springer-Verlag
Berlin Heidelberg New York London Paris Tokyo

Author

Norbert Schappacher
Max-Planck-Institut für Mathematik
Gottfried-Claren-Straße 26
5300 Bonn 3, Federal Republic of Germany

Mathematics Subject Classification (1980): Primary: 10D25, 14A20, 14K22, 12C20
Secondary: 14C99, 18F99, 14G10, 12C25, 33A15, 14K20

ISBN 3-540-18915-7 Springer-Verlag Berlin Heidelberg New York
ISBN 0-387-18915-7 Springer-Verlag New York Berlin Heidelberg

Printed in Germany

Printing and binding: Druckhaus Beltz, Hemsbach/Bergstr.
2146/3140-543210

A Rosita

qui a su débloquer la rédaction
de ce travail
janvier 1986

Налево беру и направо,
И даже, без чувства вины,
Немного у жизни лукавой
И все – у ночной тишины.

Анна Ахматова

INTRODUCTION

In two papers – n° 12 and 14 of [He], published in 1918 and 1920 – E. Hecke introduced what he called "Größencharaktere" of algebraic number fields, with a view to extending the theory of L-functions and their applications in analytic number theory. In the early 1950's the arithmetic and geometric significance of those of Hecke's characters that take algebraic values began to appear in two different, if overlapping, lines of thought. (Both of these had been anticipated in special cases by Eisenstein exactly one hundred years earlier; but none of the mathematicians working on them in the fifties seems to have been aware of their precursor at the time.) First Weil, testing a conjecture of Hasse, investigated algebraic curves over $\mathbb{Q}$ with the property that the number of $\mathbb{F}_p$ rational points on their reductions modulo p can be computed in terms of exponential sums. This led him to a study of "Jacobi sums as 'Größencharaktere' ". Secondly Deuring, developing one aspect of Weil's examples, proved that the (Hasse-Weil) L-function of an elliptic curve with complex multiplication is a (product of) Hecke L-function(s). This was then quickly generalized to higher dimensional *CM* abelian varieties by Shimura and Taniyama, with Weil providing clarification, for instance, on the Hecke characters employed in the theory.

Both approaches cover only very limited classes of algebraic Hecke characters. Jacobi sum characters were confined to cyclotomic (today: abelian) fields, and in general, not every algebraic Hecke character of such a field is given by Jacobi sums. – The product of several Hecke characters each one of which is attached to a *CM* abelian variety does no longer occur in the L-function of an abelian variety.

This last difficulty disappears in a theory of motives, as proposed by Grothendieck. There one associates with every (smooth projective) algebraic variety, in some sense, its "universal cohomology" which is an object of what Grothendieck termed a Tannakian category, and may therefore be viewed as a representation of some proalgebraic group. The product of Hecke characters then corresponds essentially to the tensor product of representations. Until the mid seventies such a category of motives existed only conjecturally; the morphisms were to be defined by the cohomology classes of algebraic correspondences, and conjectures on the existence of sufficiently many algebraic cycles had to be used to show that the construction actually yielded a Tannakian category – see [Sa]. Granting this, the semi-simplicity of the Galois action on l-adic cohomology and Tate's conjectures, one could show that a motive defined over a number field is determined up to isomorphism by its ("Hasse-Weil") L-function (defined using the étale cohomology of the motive). Consequently, two motives (which may be constructed from different varieties, but are) attached to the same Hecke character would have to be isomorphic, and in particular, would have the same periods (defined by "integrating" de Rham cohomology classes "against" Betti cohomology of the motive.)

This uniqueness principle is at the center of our work. We peruse a variety of consequences of it that can be proven, either because an analogous uniqueness principle is available in a slightly different framework – see next paragraph – or because of the special situation considered – this is the case in Chapter V. Applications include a refined version of the so called formula of Chowla and Selberg, deduced from the comparison of the motive of a basic Jacobi sum Hecke character of an imaginary quadratic field K to elliptic curves with complex multiplication by K – see Chapter III; refinements of Shimura's monomial period relations; generalizations of the formula of Chowla and Selberg to arbitrary abelian number fields – Chapter IV; and the study of motives for the theta series of Hecke characters of imaginary quadratic fields - Chapter V.

That we can actually prove theorems, not merely do an exegesis of conjectures, hinges on two insights by P. Deligne. First, he saw that one could actually construct a theory of motives by weakening the requirement on the morphisms; they no longer have to be algebraic but only "absolute Hodge" correspondences – see Chapter I, Section 2. (This idea may in fact go back to Grothendieck: see the first consequence of the Hodge conjecture discussed in §4 of [Gro].) Henceforth, when we speak of "motives", we refer to this existing theory. Second, Deligne was able to show that, on an abelian variety over $\bar{\mathbb{Q}}$, every Hodge cycle is an absolute Hodge cycle. This consequence of the Hodge conjecture provides enough absolute Hodge cycles to prove the uniqueness principle for motives of algebraic Hecke characters, within the category of motives generated by abelian varieties – see Chapter I, Theorem 5.1.

In fact, for every algebraic Hecke character of a number field K, there exists a unique motive in the category of motives over K generated by abelian varieties with potential complex multiplication. Deligne has shown around 1980 that this category is equivalent to the representations of (a subgroup of) the Taniyama group, a group scheme which had been introduced by Langlands. This structure theorem also links the motivic interpretation of Hecke characters to that proposed by Serre in [Sl] more than ten years earlier. It was also the starting point for G. Anderson's comprehensive motivic theory of Gauss and Jacobi sums, and their relations to representations of the Taniyama group, a theory which he worked out between 1982 and 1984 – see [A1] and [A2]. In Anderson's formalism, the basic observation that Fermat hypersurfaces provide motives for Jacobi sum Hecke characters of cyclotomic fields is extended to a class of characters of abelian number fields which is likely to include all sensible candidates of Hecke characters of "Jacobi sum type". We make essential use of Anderson's theory when dealing with Jacobi sum Hecke characters.

Thus, I really "take on the left and on the right" very substantial results obtained by others, and numerous little chats with many people have found their way into the "silent hours of the night" during which these pages were written.

* * *

It was my intention, in writing up the paper, to also provide a viable introduction to the background theories. More precisely, the reader should get an idea of what they are like, without however being offered complete proofs. I hope there will be readers to whom my blend of explanations and quotes appeals, and is actually helpful.

CHAPTER **0** should be completely readable for anyone with some very basic knowledge of algebraic number theory. It covers the elementary (as opposed to geometric) theory of algebraic Hecke characters, including their interpretation via Serre's groups $S_{\mathfrak{m}}$, and the definition and basic properties of Jacobi sum Hecke characters according to G. Anderson. (The Jacobi sum characters of imaginary quadratic fields are largely treated without reference to Anderson, by way of a fundamental example which is used in Chapter III.)

CHAPTER I falls into five parts.

I §1 presents the Shimura-Taniyama theory of complex multiplication of abelian varieties with a view to introducing motives for Hecke characters. The existence of the Hecke character attached to a CM abelian variety is derived using a transcendence result which implies - see [Henn] – that every semisimple abelian E-rational λ-adic representation is locally algebraic – cf. I, 1.4.

I §2 reviews the theory of motives for absolute Hodge cycles. We hope that our shortcut through this theory can serve as a reading guide for [DMOS], Chapter II, and also to the corresponding sections of [A2]. Deligne's fundamental theorem on absolute Hodge cycles of abelian varieties is only quoted from [DMOS], Chapter I, because its proof would have led us too far away from the geometric study of Hecke characters.

I §§3 - 5 cover the "naive" theory of motives for Hecke characters. In §4, a motive for every algebraic Hecke character is constructed "by hand", out of Artin motives and CM abelian varieties. Its uniqueness up to isomorphism, in the category of motives generated by abelian varieties, is derived from Deligne's theorem in §5.

I §6 treats the theory of the Taniyama group and its relation with the category of motives $\mathcal{CM}_{\mathbb{Q}}$. While in the previous sections of chapter I the reader should be able to survive with a certain knowledge of algebraic geometry, this section is deliberately sketchy. In fact, we shall make very little use of it in later chapters – except through Anderson's theory. Also, Milne is preparing a book on this subject which will also deal with Shimura varieties.

I §7 briefly reviews Anderson's theory of motives for Jacobi sum Hecke characters, and also his ulterior motives. For all the details the reader is referred to his papers.

CHAPTER II is the technical heart of this work. The formalism of the periods of motives in general and motives for Hecke characters in particular, is unfolded here. This "arithmetic linear algebra" is carried out in great generality. I am afraid this does not exactly simplify the notation and understanding of this chapter. But I do hope that this treatment of periods – which, by the way, is essentially due to Deligne – will be useful for further investigations. This chapter also contains a brief review of Deligne's rationality conjecture for special values of Hecke L-functions. This case of the conjecture is now a theorem by virtue of recent important results of Blasius [Bl] and Harder (unpublished). However, Blasius's motivic treatment of the periods c^+ is not included in my exposition because he himself applies it to questions similar to those discussed here – see his forthcoming paper [Bl′].

At the end of chapter II, after discussing the periods of Jacobi sum Hecke characters starting from the example of Fermat hypersurfaces, we deduce some relations between values of the Γ function at rational numbers which were first conjectured and proved by Deligne.

CHAPTER III is devoted to the so called formula of Chowla and Selberg. We prove a refined version of it, and show (in III §3) that it "generates" all period relations produced by Jacobi sum Hecke characters of imaginary quadratic fields. An interesting feature of the motivic treatment of this formula is that here, it is often convenient to deduce an identity of Hecke characters from an analytically accessible period relation – rather than going the other way around. – Cf. also II, 3.5.

CHAPTER IV treats Shimura's relations between periods of CM abelian varieties and generalizations of the Chowla-Selberg formula to abelian fields. The most remarkable feature here is the serious discrepancy between the potential of the method and the scarcity of information about concrete situations to which the method applies. In the Chowla-Selberg case it is often possible to determine explicitly every single character whose periods contribute to the formula. But over an arbitrary abelian field such explicit identities are usually not available, and so – in spite of the inherent precision of the method – one is led to weaken the period relations in order to get sensible statements.

Compared to the preceding chapters, CHAPTER V is really written "in short-hand". It starts by reviewing very briefly U. Jannsen's recent construction of an honest regard (absolute Hodge cycle) motive for every newform f on $\Gamma_1(N) \subset SL_2(\mathbb{Z})$ of weight ≥ 2. Then we proceed to show that this motive "lies in" $\mathcal{CM}_{\mathbb{Q}}$ if f has complex multiplication. This has to be done by hand, using Deligne's conjecture for the critical values of these modular forms.

* * *

It is a pleasure to acknowledge the hospitality of the Max-Planck-Institut für Mathematik at Bonn, where I stayed from October 1983 through January 1985. About half of this work was written there, not little influenced by Harder's interest in these questions, and his willingness to let me organize his seminar in the winter 1984 - 85 on motives for absolute Hodge cycles. Most of the suggestions explicitly acknowledged in the text I obtained through my stay in Bonn. – For the excellent typing my hearty thanks go to K. Deutler at Bonn, C. Gieseking at Göttingen.

LEITFADEN

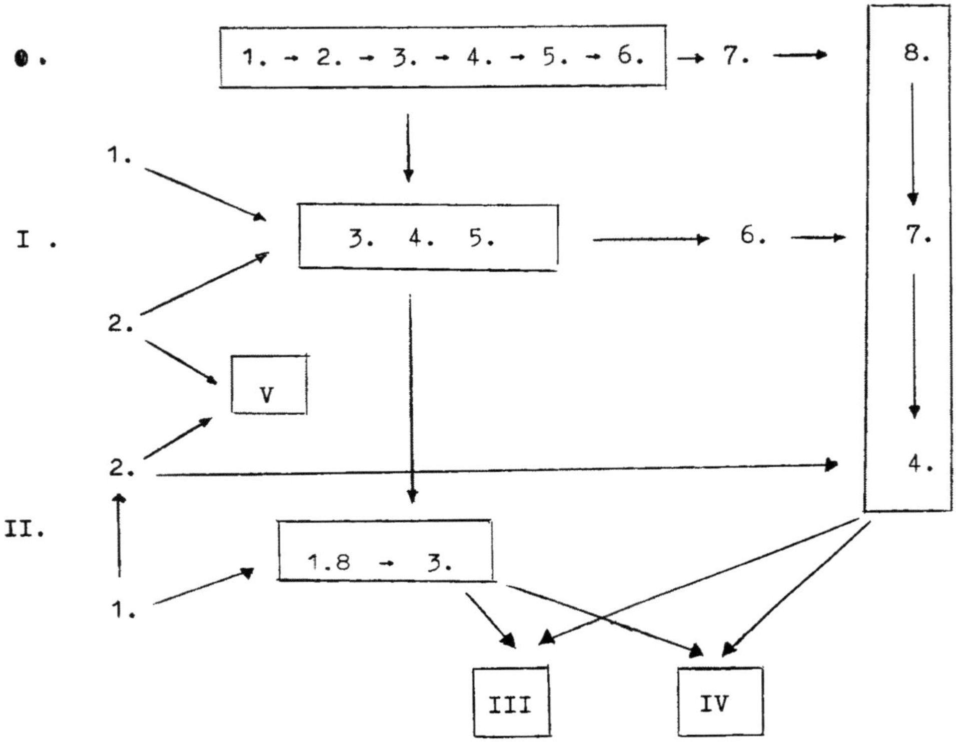

Here as well as in the internal references throughout the text, roman numerals denote chapters - chapter zero being represented by 0 - , and an expression of the form **n.m.l** , **n.m** , or **n.** refers to the corresponding formula, theorem, paragraph or section within the given chapter.

CONTENTS

CHAPTER FOUR: Abelian integrals with complex multiplication

CHAPTER FIVE: Motives of CM modular forms

CHAPTER ZERO:
Algebraic Hecke Characters

In this chapter we review the elementary theory of algebraic Hecke characters and fix some basic notation.

1. Definition

Let K and E be two number fields, i.e., finite extensions of $\mathbb{Q}$. Let $\mathfrak{f}$ be a non-zero integral ideal of K, and $T = \sum n_\sigma \sigma \in \mathbb{Z}[\mathrm{Hom}(K,\overline{E})]$ a $\mathbb{Z}$-linear combination of embeddings of K into a fixed algebraic closure $\overline{E}$ of E.

__Definition:__ [cf. [SGA $4\frac{1}{2}$], Sommes trig. § 5]: An __algebraic Hecke character__ χ __of__ K __with values in__ E, __of infinity-type__ T __and conductor dividing__ $\mathfrak{f}$, is a group homomorphism

$$\chi: I_{\mathfrak{f}} \to E^*$$

from the group $I_{\mathfrak{f}}$ of ideals of K prime to $\mathfrak{f}$ to the multiplicative group of E, such that, for any ideal $(\alpha) \in I_{\mathfrak{f}}$ generated by an $\alpha \in K^*$ with $\alpha \equiv 1 (\mathrm{mod}\ \mathfrak{f})$, and α totally positive (i.e., $\alpha^\rho > 0$ for all real embeddings $\rho: K \hookrightarrow \mathbb{R}$, symbolically: $\alpha >> 0$), one has

$$\chi((\alpha)) = \alpha^T = \prod_\sigma (\alpha^\sigma)^{n_\sigma} .$$

It is understood that, if $\mathfrak{f} | \mathfrak{f}'$, characters of conductor dividing $\mathfrak{f}$ are identified with the corresponding characters of conductor dividing $\mathfrak{f}'$ obtained by restricting to $I_{\mathfrak{f}'} \subseteq I_{\mathfrak{f}}$. The smallest $\mathfrak{f}$ (in the sense of divisibility) such that χ extends to a character of conductor dividing $\mathfrak{f}$ is called __the conductor__ of

χ, and denoted f_χ. - Note that the subgroup of ideals (α) with $\alpha \gg 0$ and $\alpha \equiv 1 \pmod f$ has finite index in I_f.

2. Algebraic homomorphisms

Recall from [SGA $4\frac{1}{2}$], Sommes trig. § 5, the various ways to view the infinity-type T of an algebraic Hecke character. In general, an <u>algebraic homomorphism</u> $t: K^* \to E^*$ is a group homomorphism such that either one of the following equivalent conditions is satisfied.

(a) For any basis $\{e_i \mid i = 1,\ldots,n\}$ of K over $\mathbb{Q}$, there is a rational function $f \in E(X_1,\ldots,X_n)$ such that

$$t(\sum a_i e_i) = f(a_1,\ldots,a_n),$$

for all $(a_i) \in \mathbb{Q}^n$.

(b) t is induced by a homomorphism of algebraic groups over E

$$R_{K/\mathbb{Q}}\, \mathbb{G}_m \times_{\mathbb{Q}} E \to \mathbb{G}_m .$$

(c) t is induced by a homomorphism of algebraic groups over $\mathbb{Q}$

$$R_{K/\mathbb{Q}}\, \mathbb{G}_m \to R_{E/\mathbb{Q}}\, \mathbb{G}_m .$$

(d) There is $T = \sum n_\sigma \sigma \in \mathbb{Z}[\mathrm{Hom}(K,\bar{E})]$ such that for all $\alpha \in K^*$,

$$t(\alpha) = \alpha^T = \prod_\sigma (\alpha^\sigma)^{n_\sigma}.$$

(e) Decompose $K \otimes_{\mathbb{Q}} E = \prod_j F_j$ (finite product of fields). There are integers m_j such that

$$t = \prod_j N_{F_j/E}^{m_j} .$$

As explained in loc. cit., the equivalence of (a) through (c) follows from elementary facts about algebraic groups, and (d), (e) are reformulations of (b) using the identification of the character group of $R_{K/\mathbb{Q}}\mathbb{G}_m$ over $\bar{E}$ with $\mathbf{Z}^{\mathrm{Hom}(K,\bar{E})}$. An analogous reformulation of (c) will be given in § 4. In the sequel we will often identify a type T like in (d) with the algebraic homomorphism t defined by it. Note that T gives rise to an algebraic homomorphism $K^* \to E^*$ if and only if $n_\sigma = n_{\tau\sigma}$, for every $\tau \in \mathrm{Gal}(\bar{E}/E)$. This is the case if T is the infinity-type of an algebraic Hecke character with values in E.

3. Infinity-types and algebraic Hecke characters

It is not true that, conversely, every algebraic homomorphism $K^* \to E^*$ occurs as infinity-type of an algebraic Hecke character of K with values in E. The first obvious constraint is that such an infinity-type has to kill all totally-positive units $\equiv 1 (\mathrm{mod}\ \mathfrak{f})$ in $\mathcal{O}_K$. As these are of finite index in $\mathcal{O}_K^*$, the proof of Dirichlet's unit theorem implies that there is an integer w such that, for any embedding $\bar{E} \hookrightarrow \mathbb{C}$, inducing an action of complex conjugation, $\sigma \mapsto \bar{\sigma}$, on $\mathrm{Hom}(K,\bar{E})$, and for any $\sigma \in \mathrm{Hom}(K,\bar{E})$, one has

$$(3.1) \qquad n_\sigma + n_{\bar{\sigma}} = w.$$

w is called the weight of T (or of χ).

Thus, for any complex conjugation of $\bar{E}$, we find

$$\chi \cdot \bar{\chi} = \mathbb{N}_{K/\mathbb{Q}}^{w} ,$$

where $\mathbb{N}_{K/\mathbb{Q}}(\mathfrak{a}) = \#(\mathcal{O}_K/\mathfrak{a})$ for an integral ideal $\mathfrak{a}$ of K. (In fact, this is true on a subgroup of finite index of $I_\mathfrak{f}$, and $\mathbb{R}_+^*$ is torsion-free.) Therefore the values of an algebraic Hecke character are pure, in the sense that all embeddings into $\mathbb{C}$ have the same absolute value. Similarly, they are what we shall call (for want of a better term) numbers of CM-type:

An algebraic number α is of CM-type if there is a (necessarily unique) conjugate α' of α such that, for all embeddings

$$\tau:\mathbb{Q}(\alpha,\alpha') \to \mathbb{C},$$

one has $\overline{\tau(\alpha)} = \tau(\alpha')$.

To make more explicit the restriction on the existence of algebraic Hecke characters imposed by the homogeneity condition $n_\sigma + n_{\bar\sigma} = \text{cst.}$, let K' be the subfield of K consisting of all $\alpha \in K$ that are of CM-type. So, K' is either totally real or a CM-field (i.e., a totally imaginary quadratic extension of a totally real field). Then in our infinity-type $T = \sum n_\sigma \cdot \sigma$, n_σ depends only on $\sigma|_{K'}$ because $n_\sigma + n_{\bar\sigma}$ is independent of the choice of complex conjugation. So one gets an element

$$T' = \sum n_{(\sigma|K')} \cdot (\sigma|K') \in \mathbb{Z}[\mathrm{Hom}\,(K',\bar{E})].$$

The fact that $n_\sigma + n_{\bar\sigma}$ is independent of σ means this:

(a) if K' is totally real, then $T' \in \mathbb{Z}\cdot\Sigma\tau$ (summed over all $\tau:K \hookrightarrow \bar{E}$);

(b) if K' is a CM-field, then T' belongs to the subgroup of $\mathbb{Z}[\mathrm{Hom}(K',\bar{E})]$ generated by the CM-types $\{\Sigma n_\sigma \sigma \mid n_\sigma \in \{0,-1\}, n_\sigma + n_{\bar\sigma} = -1\}$.

Since algebraic Hecke characters of finite order are precisely those whose infinity-type is trivial, we see that

(a) if K' is totally real, then every algebraic Hecke character χ of K is of the form $\chi = \mu\cdot\mathbb{N}_{K/\mathbb{Q}}^{w/2}$, where μ is of finite order and $w \in 2\mathbb{Z}$.

(b) if K' is a CM-field, we have $\chi = \mu\cdot(\varphi \circ N_{K/K'})$ for some algebraic Hecke character φ of K', and μ a character of finite order of K.

Consequently, the field of values of an algebraic Hecke character is either $\mathbf{Q}$ or a CM-field.

Every algebraic homomorphism $T = \Sigma n_\sigma \cdot \sigma : K^* \to E^*$ which satisfies the homogeneity condition $n_\sigma + n_{\overline{\sigma}} =$ cst. as above is the infinity-type of some algebraic Hecke character χ of K with values in a finite extension of E. The construction of χ is straightforward: Choose an ideal $\mathfrak{f}$ such that

$$\{\varepsilon \in o_K{}^* \mid \varepsilon >> 0 \text{ and } \varepsilon \equiv 1(\mathrm{mod}\,\mathfrak{f})\}^T = 1,$$

and take any extension of

$$\chi|\{(\alpha) \mid \alpha >> 0 \text{ and } \alpha \equiv 1(\mathrm{mod}\ \mathfrak{f})\}$$

to all of $I_\mathfrak{f}$ (cf.,e.g., [Shim], Lemma 7.45). Since roots of values of T are extracted in this process one cannot do without extending E, in general. If we insist on keeping E, we may pass to $T \circ N_{L/K}$, for a suitable finite (abelian) extension L of K. Fixing E and K, all one can assert in general is that $m \cdot T$, for a suitable $m \in \mathbb{N}$, will come from an algebraic Hecke character of K with values in E. (Cf. [DP], 8.2.) - We shall not be concerned with the problem of bounding the conductor of the character χ which we have just shown to exist. For this, cf. [Schm],I.2.

4. The Hodge decomposition

A homomorphism $R_{K/\mathbb{Q}}\mathbb{G}_m \to R_{E/\mathbb{Q}}\mathbb{G}_m$ over $\overline{E}$ (cf. §2,(c)) is a system of characters of $R_{K/\mathbb{Q}}\mathbb{G}_m$ indexed by $\mathrm{Hom}(E,\overline{E})$. This yields a description of infinity types whose relation with the title of this section will become apparent in the next chapter (see in particular I, 1.7,4.2,6.1.5; cf. [DP], 8.2).

Let $t: K^* \to E^*$ be an algebraic homomorphism, and $\tau \in \mathrm{Hom}(E,\mathbb{C})$. Then $\tau \circ t: K^* \to (E^\tau)^*$ is again an algebraic homomorphism whose type will be written

$$T_\tau = \sum n(\sigma,\tau) \cdot \sigma ,$$

where σ now ranges over all embeddings of K into the algebraic closure of $\mathbb{Q}$ in $\mathbb{C}$, or simply $\sigma: K \to \mathbb{C}$.

If $T = \sum_{\eta:K\to\bar{E}} n_\eta \cdot \eta$ is the type of t, then $n_\eta = n(\sigma,\tau)$ if and only if $\sigma = \tilde{\tau}\circ\eta$, for any extension $\tilde{\tau}$ of τ to $\bar{E}$.

We have $n(\sigma,\tau) = n(\alpha\sigma,\alpha\tau)$ for all $\alpha \in \mathrm{Aut}\,\mathbb{C}$ because $t: R_{K/\mathbb{Q}}\mathbb{G}_m \to R_{E/\mathbb{Q}}\mathbb{G}_m$ is defined over $\mathbb{Q}$ (cf. I,6.1.2). Furthermore, if t is the infinity-type of an algebraic Hecke character χ of weight w, then

$$w = n(\sigma,\tau) + n(c\sigma,\tau) = n(\sigma,\tau) + n(\sigma,c\tau),$$

for any $\sigma \in \mathrm{Hom}(K,\mathbb{C})$, $\tau \in \mathrm{Hom}(E,\mathbb{C})$, where c = complex conjugation.

5. Adèles

Algebraic Hecke characters may, of course, be read on the idèles $K^*_{\mathbb{A}}$ of K: cf. [W],[1955c] where algebraic Hecke characters were introduced as characters of the idèle class group "of type (A_0)".

First, given χ as in § 1, there clearly exists a unique group homomorphism

$$\chi_{\mathbb{A}}: K_{\mathbb{A}}^* \to E^*$$

such that

(a) $\chi_{\mathbb{A}}^{-1}(1)$ is open in $K^*_{\mathbb{A}}$;
(b) $\chi_{\mathbb{A}|K^*} = T: K^* \to E^*$;
(c) $\chi_{\mathbb{A}}(\pi_{\mathfrak{p}}) = \chi(\mathfrak{p})$, for all prime ideals $\mathfrak{p}$ of K not dividing $\mathfrak{f}$.

Here, $\pi_{\mathfrak{p}}$ denotes any idèle having a uniformizing parameter at $\mathfrak{p}$, and 1 at all other components.

Since $\chi_{\mathbb{A}}$ takes values in E^*, it could not be an idèle class character, and its restrictions to individual completions of K are not very interesting. But this can be changed by conveniently "localizing over E":

Being an algebraic homomorphism, T induces a continuous homomorphism $K^*_{\mathbb{A}} \to E_{\mathbb{A}}^*$ - see, e.g., condition (c) of § 2. Given any place λ of E, denote by T_λ the composite with projection onto the λ-component of $E_{\mathbb{A}}$:

$$T_\lambda : K^*_{\mathbb{A}} \xrightarrow{T} E^*_{\mathbb{A}} \to E^*_\lambda ,$$

and write

$$\chi_\lambda = \chi_{\mathbb{A}} \cdot T_\lambda^{-1} : K^*_{\mathbb{A}} \to E^*_\lambda .$$

Then χ_λ is an idèle class character, i.e., a continuous homomorphism $K^*_{\mathbb{A}}/K^* \to E^*_\lambda$.

If λ is a finite place, E_λ is a totally disconnected topological space, so $\ker \chi_\lambda$ contains the connected component of 1 in $K^*_{\mathbb{A}}$. By class-field-theory, χ_λ factorizes: $\mathrm{Gal}(K^{ab}/K) \to E^*_\lambda$ as the 1-dimensional λ-adic Galois representation with

$$\chi_\lambda(\mathrm{Frob}\ \mathfrak{p}) = \chi(\mathfrak{p}) \in E^* \hookrightarrow E^*_\lambda ,$$

for any prime ideal $\mathfrak{p}$ of K not dividing $\mathfrak{f}.\ \mathbb{N}\lambda$. Here, Frob $\mathfrak{p}$ is a "geometric Frobenius" at $\mathfrak{p}$, i.e., we normalize the reciprocity map of class-field-theory to be the reciprocal of the Artin map. This is done to comply with [DP]. Note that the "eigenvalues" of Frob $\mathfrak{p}$ with respect to this λ-adic representation are purely of absolute value $(\mathbb{N}\mathfrak{p})^{w/2}$.

If λ is a complex place of E, we get two (possibly equal) continuous homomorphisms, or "quasi-characters of the idèle class group" in the sense of [Tt] or [W3], chap. VII:

$${}_1\chi_\lambda, {}_2\chi_\lambda : K^*_{\mathbb{A}}/K^* \to E^*_\lambda \overset{\longrightarrow}{\underset{\longrightarrow}{\sim}} \mathbb{C}^* ,$$

according to the two continuous isomorphisms $E_\lambda \cong \mathbb{C}$.

If λ is a real place, there is just one such character $\chi_\lambda : K^*_{\mathbb{A}}/K^* \to \mathbb{R}^* \hookrightarrow \mathbb{C}^*$.

In another language, we get for each infinite place λ, one or two automorphic forms on $GL(1, K_{\mathbb{A}})$.

6. L-functions

To every complex embedding $\tau : E \to \mathbb{C}$ is attached the "Größencharakter" (in Hecke's sense) $\tau \circ \chi$. If τ induces the infinite place λ of E, then $\tau \circ \chi$ corresponds to (one of) the idèle class character(s) ${}_{(j)}\chi_\lambda$. Consider the Hecke L-function

$$L(\chi^\tau, s) = \sum_{\substack{(\mathfrak{a}, \mathfrak{f}_\chi) = 1 \\ \mathfrak{a} \subset \mathfrak{o}_K}} (\tau \circ \chi)(\mathfrak{a}) \cdot \mathbb{N}\,\mathfrak{a}^{-s} = \prod_{\mathfrak{p} \nmid \mathfrak{f}_\chi} (1 - (\tau \circ \chi)(\mathfrak{p}) \cdot \mathbb{N}\mathfrak{p}^{-s})^{-}$$

$$(\text{for } \operatorname{Re}(s) > \tfrac{w}{2} + 1).$$

We write formally

$$L^*(\chi, s) = (L(\chi^\tau, s))_{\tau: E \to \mathbb{C}}\ ,$$

so that $L^*(\chi,s)$ is an array of L-functions taking values in $\mathbb{C}^{\operatorname{Hom}(E,\mathbb{C})} = E \otimes_{\mathbb{Q}} \mathbb{C}$.

Recall the general form of the functional equation of the $L(\chi^\tau, s)$ - cf. [He], p. 272 f; [Tt] or [W3], VII-7. Put

$$\Gamma_{\mathbb{R}}(s) = \pi^{-\frac{s}{2}}\, \Gamma(\tfrac{s}{2})\, ; \quad \Gamma_{\mathbb{C}}(s) = \Gamma_{\mathbb{R}}(s)\Gamma_{\mathbb{R}}(s+1) = 2(2\pi)^{-s}\Gamma(s).$$

<u>For a real place</u> v of K (whose existence implies, in the notation of § 4, that all $n(\sigma,\tau)$ are equal to $\frac{w}{2}$), put

$$L_v(\chi^\tau, s) = \Gamma_{\mathbb{R}}(s + \varepsilon - \tfrac{w}{2}),$$

where $\varepsilon = 0$ or 1, such that the v-component $\chi_v^\tau : K_v^* \to \mathbb{C}^*$ of χ_λ satisfies $\chi_v^\tau(-1) = (-1)^{\varepsilon + w/2}$. <u>For a complex place</u> v of K, corresponding to the pair $\sigma, \bar{\sigma}: K \to \mathbb{C}$ of complex embeddings of K, put

$$L_v(\chi^\tau, s) = \Gamma_{\mathbb{C}}(s - \inf(n(\sigma,\tau), n(\bar{\sigma},\tau))).$$

Then, setting

$$\Lambda(\chi^\tau, s) = \prod_{v \mid \infty} L_v(\chi^\tau, s) \cdot L(\chi^\tau, s),$$

we get a meromorphic continuation to the whole complex plane with functional equation of the type

$$\Lambda(\chi^\tau,s) = \varepsilon(\chi^\tau,s) \cdot \Lambda((\chi^\tau)^{-1}, 1-s),$$

where $\varepsilon(\chi^\tau,s) = W(\chi^\tau)\cdot\{|d_K| \cdot \mathbb{N}f_\chi\}^{1/2-s}$, for some constant $W(\chi^\tau)$ of absolute value 1, and d_K the discriminant of K (over $\mathbb{Q}$). - As $\overline{\chi^\tau} = \mathbb{N}^w\cdot(\chi^\tau)^{-1}$, this functional equation may be rewritten as one relating $L(\chi^\tau,s)$ to $L(\overline{\chi^\tau},w+1-s)$.

7. Serre's group

In [Sℓ], chap. II, Serre has given an interpretation of algebraic Hecke characters which generalizes the definitions (b) or (c) of algebraic homomorphisms recalled in § 2.

7.1 Put:

$$U_{\mathbb{A},f} = \left\{ (x_v) \in K^*_{\mathbb{A}} \;\middle|\; \begin{array}{l} x_v > 0 \quad \text{if} \quad v \quad \text{is real} \\ x_v \equiv 1 \,(\mathrm{mod}\; f_v) \quad \text{if} \quad v|f \\ x_v \in \mathfrak{o}_v^* \quad \text{if} \quad v \quad \text{is finite} \end{array} \right\}.$$

Then $\chi_{\mathbb{A}}$ factorizes through $K^*_{\mathbb{A}}/U_{\mathbb{A},f}$, and we have the diagram

$$\begin{array}{ccccccccc} 1 & \to & K^*/U_f & \to & K^*_{\mathbb{A}}/U_{\mathbb{A},f} & \to & C_f & \to & 1 \\ & & {\scriptstyle T_{\mathbb{A}}} \searrow & & \swarrow {\scriptstyle \chi_{\mathbb{A}}} & & & & \\ & & & E^* & & & & & \end{array}$$

where $U_f = U_{\mathbb{A},f} \cap K^*$, and C_f is the ray class group of K mod f. Recall that C_f is a finite abelian group. Now, K^*/U_f is the group of $\mathbb{Q}$-rational points of the $\mathbb{Q}$-torus

$$Z_{K,f} = (R_{K/\mathbb{Q}}\,\mathbb{G}_m)/\overline{\Gamma_f},$$

where $\overline{\Gamma_f}$ is the Zariski-closure of a suitable arithmetic subgroup Γ_f of $R_{K/\mathbb{Q}}\,\mathbb{G}_m$. Serre shows how to construct a $\mathbb{Q}$-algebraic group $S_{K,f}$ of multiplicative type (i.e., $S_{K,f}$

is the product of a torus by a finite abelian group) which is an extension of C_f by $Z_{K,f}$ such that $S_{K,f}(\mathbb{Q}) = K^*_{\mathbb{A}}/U_{\mathbb{A},f}$. In fact, define $S_{K,f}$ via its character group:

$$X(S_{K,f}) = \mathrm{Hom}(S_{K,f} \times_{\mathbb{Q}} \overline{\mathbb{Q}}, \mathbb{G}_m/\overline{\mathbb{Q}}) =$$

$$= \left\{ (\eta,\xi) \;\middle|\; \begin{array}{l} \eta : K^*_{\mathbb{A}}/U_{\mathbb{A},f} \to \overline{\mathbb{Q}}^* \text{ homomorphism} \\ \xi \in \mathrm{Hom}(Z_{K,f} \times_{\mathbb{Q}} \overline{\mathbb{Q}}, \mathbb{G}_m/\overline{\mathbb{Q}}) \\ \underline{\text{and}} \quad \eta = \xi \text{ on } K^*/U_f \end{array} \right\}$$

(as $\mathrm{Gal}(\overline{\mathbb{Q}}|\mathbb{Q})$-module).

__7.2__ An algebraic Hecke character of K with values in E can then be viewed as a __representation defined over__ E __of algebraic groups__

$$S_K = \varprojlim_{f} S_{K,f} \to GL(1) ,$$

or equivalently, as a homomorphism of algebraic groups defined over $\mathbb{Q}$:

$$S_K \to R_{E/\mathbb{Q}} \mathbb{G}_m .$$

__7.3__ S_K sits in the exact sequence of $\mathbb{Q}$-algebraic groups (obtained as projective limit over f):

$$(7.3.1) \qquad 1 \to Z_K \to S_K \to \mathrm{Gal}(K^{ab}/K) \to 0,$$

where Z_K can be described as follows.

__7.3.2__ Given an algebraic Hecke character of K with values in E, as a representation $S_K \xrightarrow{\chi} R_{E/\mathbb{Q}} \mathbb{G}_m$, its infinity-type is obtained simply by restricting to Z_K: $Z_K \xrightarrow{T} R_{E/\mathbb{Q}} \mathbb{G}_m$ (cf. §4) and Z_K is the largest quotient of $R_{K/\mathbb{Q}} \mathbb{G}_m$ through which all infinity-types of algebraic Hecke characters of K factorize.

Let $\overline{\mathbb{Q}}$ be the algebraic closure of $\mathbb{Q}$ in $\mathbb{C}$, and consider temporarily all fields K as subfields of $\overline{\mathbb{Q}}$. Write $\tau : E \hookrightarrow \overline{\mathbb{Q}} \subset \mathbb{C}$. Then $\tau \circ T$ can be interpreted as a character $T_\tau : Z_K \to \mathbb{G}_m$ over $\overline{\mathbb{Q}}$. Denoting complex conjugation on $\overline{\mathbb{Q}}$ by c, and by

$G_{CM} \subset Gal(\bar{\mathbb{Q}}/\mathbb{Q})$ the subgroup fixing all algebraic numbers of CM-type, it follows that the character group of Z_K is given by:

$$X(Z_K) = \left\{ \lambda \in X(R_{K|\mathbb{Q}}\mathbb{G}_m) \;\middle|\; \begin{array}{l} \lambda^s = \lambda, \text{ for all } s \in G_{CM} \\ \lambda(c\sigma) + \lambda(\sigma) \text{ indep. of } \sigma \end{array} \right\},$$

where we identify $X(R_{K|\mathbb{Q}}\mathbb{G}_m) = \mathbb{Z}[\mathrm{Hom}(K,\bar{\mathbb{Q}})]$, and define $\lambda^s(\sigma) = \lambda(s^{-1}\circ\sigma)$, for $s \in Gal(\bar{\mathbb{Q}}/\mathbb{Q}), \sigma \in \mathrm{Hom}(K,\bar{\mathbb{Q}})$.

7.3.3 Consequently, if $K' \subset K$ is the field of numbers of CM-type in K, then $Z_K = Z_{K'}$, and

$$Z_{K'} = \begin{cases} Z_{\mathbb{Q}} = \mathbb{G}_m/\mathbb{Q}, & \text{if } K' \text{ is totally real} \\ R_{K'/\mathbb{Q}}\mathbb{G}_m/\ker(N_{K'/K_0'}: R_{K'/\mathbb{Q}}\mathbb{G}_m \to R_{K_0'/\mathbb{Q}}\mathbb{G}_m), & \text{if } K' \text{ is a CM-field with } K_0' \text{ as maximal totally real subfield.} \end{cases}$$

In particular, for $K_1 \subset K_2$, the norm maps

$$N_{K_2/K_1} : R_{K_2/\mathbb{Q}}\mathbb{G}_m \to R_{K_1/\mathbb{Q}}\mathbb{G}_m$$

factor through $Z_{K_2} \to Z_{K_1}$, allowing us to define

$$Z = \varprojlim_{K} Z_K = \varprojlim_{K \text{ CM-field}} Z_K .$$

The infinity-types of all algebraic Hecke characters can be regarded as characters of Z (identifying T_τ on $R_{K_1/\mathbb{Q}}\mathbb{G}_m$ with $T_\tau \circ N_{K_2/K_1}$):

$$X(Z) = \left\{ f: Gal(\bar{\mathbb{Q}}/\mathbb{Q}) \to \mathbb{Z} \;\middle|\; \begin{array}{l} f \text{ locally constant} \\ f^s = f, \text{ for all } s \in G_{CM} \\ f(c\sigma) + f(\sigma) \text{ indep. of } \\ \sigma \in Gal(\bar{\mathbb{Q}}/\mathbb{Q}) \end{array} \right\}$$

7.3.4 Thus, some invariants of χ can be viewed as homomorphisms of (pro-) algebraic groups. E.g.,

$X(Z) \to \mathbb{Z} = X(\mathbb{G}_m)$ — <u>gives rise to:</u>

$f \to f(1) + f(c)\ (=w)$ — $\tilde{w}: \mathbb{G}_m \to Z\ (/\mathbb{Q})$

$f \to f(1)\ \ (= n(\sigma,\tau))$ — $\tilde{\mu}: \mathbb{G}_m \to Z\ (/\mathbb{C})$

<u>7.4</u> The sequence 7.3.1 admits a <u>natural section over the finite idèles</u> $\mathbb{Q}^*_{\mathbb{A}}f$ whose construction is reminiscent of the way in which we passed from $\chi_{\mathbb{A}}$ to χ_λ, for a finite place λ of E, in § 5. As $S_K(\mathbb{Q}) = \varprojlim_f K^*_{\mathbb{A}}/U_{\mathbb{A},f}$, there is a natural continuous map

$$f: K^*_{\mathbb{A}} \to S_K(\mathbb{Q}) \hookrightarrow S_K(\mathbb{Q}_{\mathbb{A}}f)\ .$$

On the other hand, $Z_K(\mathbb{Q}_{\mathbb{A}}f)$ is also a quotient of $K^*_{\mathbb{A}}$, whence a continuous map

$$g: K^*_{\mathbb{A}} \to Z_K(\mathbb{Q}_{\mathbb{A}}f) \to S_K(\mathbb{Q}_{\mathbb{A}}f).$$

f and g obviously agree on K^*, and as $S_K(\mathbb{Q}_{\mathbb{A}}f)$ is a totally disconnected topological space, the quotient f/g factors through a continuous homomorphism

$$\varepsilon: \mathrm{Gal}(K^{ab}/K) \to S_K(\mathbb{Q}_{\mathbb{A}}f)$$

which is the section sought.

Given an algebraic Hecke character as a homomorphism of $\mathbb{Q}$-algebraic groups

$$\chi: S_K \to R_{E/\mathbb{Q}}\, \mathbb{G}_m \quad ,$$

we can recover $\chi_{\mathbb{A}}$ as the map induced by χ on the $\mathbb{Q}$-rational points of S_K, $R_{E/\mathbb{Q}}\,\mathbb{G}_m$. As for χ_λ, for λ a finite place of E, it is the λ-component of

$$\mathrm{Gal}(K^{ab}/K) \overset{\varepsilon}{\to} S_K(\mathbb{Q}_{\mathbb{A}}f) \xrightarrow{\chi/\mathbb{Q}_{\mathbb{A}}f} R_{E/\mathbb{Q}}\mathbb{G}_m(\mathbb{Q}_{\mathbb{A}}f) = E^*_{\mathbb{A}}f.$$

It is obvious how to mimick the construction of fg^{-1} at the infinite place of $\mathbb{Q}$. The result will no longer factor through $\mathrm{Gal}(K^{ab}/K)$, but gives the characters χ_λ, for $\lambda|\infty$, introduced in § 5.

8. Jacobi sum Hecke characters

8.0.1 History

Although special cases are already present in Eisenstein [Ei 3] the notion of Gauss or Jacobi sums viewed as Hecke characters really starts with André Weil: [W II], 1952d, for the cyclotomic case; [W III], 1974d, over abelian number fields. Cf. also the beautiful [W III], 1974c. Several authors have then extended the class of characters amenable to Weil's method - see [Kb],[KL],[Li] -, and proved results about special values of their L-functions (to wit, special cases of Lichtenbaum's "Γ-hypothesis"): [Br],[BL],[Li].

On the other hand, a thoroughly geometric study of Weil's Jacobi sum Hecke characters - with a view to majorize exponential sums - was done by Deligne in [SGA $4\frac{1}{2}$], Sommes trig. - The "motivic" picture of Jacobi sum characters over cyclotomic fields is implicitly discussed in [DMOS], I § 7.

Recently, G. Anderson took up the subject introducing, in [A2], a very smooth and efficient formalism as well as a geometric interpretation for a class of Jacobi sum Hecke characters which includes all Hecke characters of abelian fields that have ever been proposed as candidates of Jacobi sum Hecke characters. More precisely, I checked that Anderson's class coincides with the one defined by Kubert,[Kb]. The geometric interpretation makes it seem very unlikely that new reasonable candidates for Jacobi sum Hecke characters of abelian fields can be proposed. Anderson's work (in fact, essentially already the earlier [A1]) definitely links up the "Γ-hypothesis" with Deligne's rationality conjecture of [DP]. This has actually been the starting point of the present work - see [GS'], announcement made after Cor. 1.2.

8.0.2 In this section we shall, as of 8.2, introduce Anderson's class of Jacobi sum Hecke characters and briefly discuss, in 8.4, the corresponding notion of Stickelberger ideal (of an abelian number field) - which is easily seen to coincide with Sinnott's ,[Sin]. - Our account of [A2] will continue in I § 7, where we describe Anderson's motives for Jacobi sum Hecke characters - also touching upon his 'ulterior motives' -, and will be concluded in II § 4, with the calculation of their periods in terms of values of the Γ-function at rational numbers. - Proofs will often be replaced by a reference.

To make things more concrete we begin, in 8.1, with a family of examples of Jacobi sum Hecke characters (all included already in [W III], 1974d) which will play a prominent role in chapter III.

8.1 The basic Jacobi sum character of an imaginary quadratic number field.

Let $K = \mathbb{Q}(\sqrt{-D})$ be the imaginary quadratic number field of discriminant $-D < -8$. (The exceptional cases $D = 3,4,8$ will be treated in 8.3.2.) Pick an embedding

$$K \hookrightarrow L = \mathbb{Q}(\mu_D) = \mathbb{Q}(e^{2\pi i/D}) \subset \mathbb{C} .$$

K is the unique quadratic subfield of L. For a prime $\mathfrak{P}$ of L not dividing D write $L(\mathfrak{P}) = \mathbb{Z}[\mu_D]/\mathfrak{P}$ the residue field and $\chi_{D,\mathfrak{P}}$ the D-th power residue symbol modulo $\mathfrak{P}$: for $x \in L(\mathfrak{P})$,

$$\chi_{D,\mathfrak{P}}(x) \in \mu_D \cup \{0\} \quad \text{and} \quad \chi_{D,\mathfrak{P}}(x) \pmod{\mathfrak{P}} = x^{\frac{\mathbb{N}\mathfrak{P}-1}{D}} .$$

Finally, put $e(z) = \exp(2\pi i\, z)$, and denote by tr the trace map from $L(\mathfrak{P})$ down to its prime field $\mathbb{F}_p$. The basic Jacobi sum character of K is given by its values on prime ideals $\mathfrak{p} \nmid D$ of K as follows.

$$(8.1.1) \qquad J_D(\mathfrak{p}) = \prod_{\mathfrak{P}|\mathfrak{p}} \Big(-\sum_{x \in L(\mathfrak{P})} \chi_{D,\mathfrak{P}}(x)\; e((\mathrm{tr}\, x)/p)\Big),$$

where $\mathfrak{P}$ runs over the primes of L dividing $\mathfrak{p}$. Extending multiplicatively to the group $I_{(D)}$ of all ideals prime to D in K, this gives a homomorphism

$$I_{(D)} \to K^* ,$$

as is easily seen from the behaviour of Gauss sums under conjugation: see [W III], 1974c, § 1. But it is by no means obvious, a priori, that J_D is a Hecke character, i.e., that it "admits a conductor" $\mathfrak{f}$, as in § 1. Suppose we knew this. Then Stickelberger's theorem would give us the infinity type of J_D, as follows. Identify as usual

$$(\mathbb{Z}/D\mathbb{Z})^* \xrightarrow{\sim} \mathrm{Gal}\,(L/\mathbb{Q})$$
$$\alpha \longrightarrow \sigma_\alpha : \zeta \to \zeta^\alpha \qquad (\zeta \in \mu_D) \quad .$$

Write the Dirichlet character corresponding to K as

(8.1.2)
$$\varepsilon : (\mathbb{Z}/D\mathbb{Z})^* \longrightarrow \{\pm 1\}$$
$$\varepsilon(\alpha) = \begin{cases} 1 & \text{if} \quad \sigma_\alpha|_K = 1 \\ -1 & \text{if} \quad \sigma_\alpha|_K = c \ \text{(complex conjugation)} \end{cases}$$

Lift ε back to $\mathbb{Z}$ when convenient, also extending it to numbers not prime to D by 0. Thus $\varepsilon(p) = (\frac{-D}{p})$ (Legendre's symbol), for all rational primes p. Define, for α running over $(\mathbb{Z}/D\mathbb{Z})^*$

(8.1.3) $$n_1 = \sum_{\varepsilon(\alpha) = -1} \langle\tfrac{\alpha}{D}\rangle \quad ; \ n_c = \sum_{\varepsilon(\alpha)=+1} \langle\tfrac{\alpha}{D}\rangle \quad ,$$

where

(8.1.4) $\langle . \rangle : \mathbb{Q}/\mathbb{Z} \to \mathbb{Q}$ is the representative in $[0,1)$ of a class mod $\mathbb{Z}$.

Then an easy calculation, starting, e.g., from [W III], 1974c, § 15, shows that

$$J_D(\mathfrak{p})\cdot o_K = \mathfrak{p}^{(n_1 1 + n_c c)}$$

Now, the trick of Gauss as a young man, and the analytic class number formula of Dirichlet give the two equations (remember that $D \neq 3,4$):

$$n_1 + n_c = \varphi(D)/2$$

$$n_1 - n_c = h_D \quad ,$$

where $\varphi(D) = \#(\mathbb{Z}/D\mathbb{Z})^*$ is Euler's φ-function, and h_D is the class number of K. Therefore the infinity type of J_D would have to be

$$(8.1.5) \qquad T_D = \frac{1}{2}[(\frac{\varphi(D)}{2} + h_D) \cdot 1 + (\frac{\varphi(D)}{2} - h_D) \cdot c] .$$

This does give an algebraic homomorphism of K^* into itself because, by genus theory , $\frac{\varphi(D)}{2} \equiv h_D \pmod 2$ - this is why we had to exclude $D = 8$ also!

It is proved in [W III], 1974d, that J_D is actually a Hecke character of K , with defining ideal $\mathfrak{f}$ dividing a power of D. Alternatively, this follows from Anderson's interpretation: see I § 7.

8.2 Anderson's formalism

The latent reference for this subsection is [A2], § 2.

8.2.1 Let $\mathbb{B}$ be the free abelian group on $\mathbb{Q}/\mathbb{Z}\setminus\{0\}$. For $\underline{a} = \Sigma n_a[a] \in \mathbb{B}$,let $m(\underline{a})$ be the order of the subgroup of $\mathbb{Q}/\mathbb{Z}$ generated by $\{a \in \mathbb{Q}/\mathbb{Z} \mid n_a \neq 0\}$. Extend the function 8.1.4 to $\mathbb{B}$ by the rule

$$\langle\underline{a}\rangle = \sum n_a \langle a\rangle.$$

Let $\overline{\mathbb{Q}}$ be the algebraic closure of $\mathbb{Q}$ in $\mathbb{C}$, and let Gal $(\overline{\mathbb{Q}}/\mathbb{Q})$ act on $\mathbb{B}$ via its action on roots of 1: Writing $\Psi: \mathrm{Gal}(\overline{\mathbb{Q}}/\mathbb{Q}) \to \hat{\mathbb{Z}}^*$ the cyclotomic character defined by $\zeta^s = \zeta^{\Psi(s)}$, for all $s \in \mathrm{Gal}(\overline{\mathbb{Q}}/\mathbb{Q})$ and $\zeta \in \overline{\mathbb{Q}}^*$ a root of 1, we set

$$\underline{a}^s = (\sum n_a[a])^s = \sum n_a \; [\Psi(s)a] \quad .$$

Given a number field $K \subset \overline{\mathbb{Q}}$, write $\mathbb{B}_K = \mathbb{B}^{G(\overline{\mathbb{Q}}/K)}$ the subgroup of elements invariant under $\mathrm{Gal}(\overline{\mathbb{Q}}/K)$. Given K and $\underline{a} \in \mathbb{B}_K$, define

$$\theta_K(\underline{a}) : G(\overline{\mathbb{Q}}/\mathbb{Q})/G(\overline{\mathbb{Q}}/K) \to \mathbb{Q}$$

$$\sigma \mapsto \langle \sigma^{-1} \underline{a} \rangle .$$

In the application, K will be abelian over $\mathbb{Q}$ and θ_K will be read on $Gal(K/\mathbb{Q})$.

Also let $\mathbb{B}^0 = \{\underline{a} \in \mathbb{B} \mid \sum n_a a = 0 \text{ in } \mathbb{Q}/\mathbb{Z}\}$, and
$\mathbb{B}_K^0 = \mathbb{B}^0 \cap \mathbb{B}_K$.

<u>8.2.2</u> Let p be a rational prime and let $\mathbb{B}_{(p)}$ be the subgroup of $\mathbb{B}$ generated by elements of the form

$$\sum_{j=1}^{f} [p^j a] ,$$

where f is a positive integer and $0 \neq a \in \mathbb{Q}/\mathbb{Z}$ is such that $(p^f - 1)a = 0$.

We assume that, for every rational prime p, an extension of the p-adic absolute value $|\,|_p$ to $\overline{\mathbb{Q}}$ has been chosen. So, in any number field $L (\subset \overline{\mathbb{Q}})$, there is a privileged prime divisor $\mathfrak{P}$ of p.

<u>There is a unique homomorphism</u>

$$g_p : \mathbb{B}_{(p)} \to \overline{\mathbb{Q}}^*$$

<u>such that, for all integral powers</u> $q = p^f \geq p$, <u>and all</u> $0 \neq a \in \mathbb{Q}/\mathbb{Z}$ <u>with</u> $(q-1)a = 0$, <u>one has</u>

$$g_p\left(\sum_{j=1}^{f} [p^j a]\right) = - \sum_{\zeta^{q-1}=1} \zeta^{-\langle a \rangle (q-1)} \cdot \mathfrak{e}\left(\frac{t(q,\zeta)}{p}\right)$$

<u>with</u> $t(q,\zeta) \in \mathbb{Z}$ <u>and</u> $t(q,\zeta) \equiv \sum_{j=1}^{f} \zeta^{p^j} \pmod{\mathfrak{P}}$, <u>for</u> $\mathfrak{P}$ <u>the chosen prime over</u> p <u>in the field</u> $\mathbb{Q}(\mu_{q-1})$.
This is Anderson's version of the theorem of Hasse and Davenport. - Note the - in the exponent of ζ !

8.2.3 Let $K \subset \overline{\mathbb{Q}}$ be a number field which is abelian over $\mathbb{Q}$. Let $\underline{a} \in \mathbb{B}_K$ and $\mathfrak{p}$ a prime ideal of K with $\mathfrak{p} \nmid m(\underline{a})$. Call p the rational prime below $\mathfrak{p}$, and write $D(K,\mathfrak{p}) \subset \mathrm{Gal}\,(\overline{\mathbb{Q}}/\mathbb{Q})$ the open subset of all s such that $\mathfrak{p}^s$ is the privileged prime above p in K. Thus $D(K,\mathfrak{p})$ consists of full left cosets of the decomposition group $D(p)$ of $|\,|_p$, as well as of $G(\overline{\mathbb{Q}}/K)$.

Put

$$g_K(\underline{a},\mathfrak{p}) = g_p\Big(\sum_{\sigma \in D(K,\mathfrak{p})/G(\overline{\mathbb{Q}}/K)} \sigma^{-1}\underline{a} \Big) ,$$

where g_p was defined in 8.2.2: To see that $\sum \sigma^{-1}\underline{a}$ actually lies in $\mathbb{B}_{(p)}$, note that

$$\mathbb{B}_{(p)} = \mathbb{B}^{D(p)} \cap \{\underline{a} \in \mathbb{B} \mid p \nmid m(\underline{a})\} .$$

8.2.4 A Jacobi sum Hecke character (according to Anderson) is a character of the form $J_K(\underline{a})$, where:

- K is an abelian number field, $K \subset \overline{\mathbb{Q}}$,
- $\underline{a} \in \mathbb{B}_K^0$,

and $J_K(\underline{a})$ is given on prime ideals $\mathfrak{p}$ of K not dividing $m(\underline{a})$ by the rule

- $J_K(\underline{a})(\mathfrak{p}) = g_K(\underline{a},\mathfrak{p})$.

The fact that $J_K(\underline{a})$ is actually a Hecke character of K, with defining ideal dividing a power of $m(\underline{a})$, hinges on the condition $\underline{a} \in \mathbb{B}^0$, and can either be dug out of [Kb], or derived from Anderson's geometric interpretation: see I § 7.

8.2.5 Elementary properties of Gauss sums imply that $J_K(\underline{a})$ is galois equivariant:

$$[J_K(\underline{a})(\mathfrak{a})]^s = J_K(\underline{a})(\mathfrak{a}^s) = J_K(\underline{a}^s)(\mathfrak{a}) ,$$

for all $s \in \mathrm{Gal}(\overline{\mathbb{Q}}/\mathbb{Q})$ and any ideal $\mathfrak{a}$ of K prime to $m(\underline{a})$. In particular, $J_K(\underline{a})$ takes values in K^*.

8.2.6 It is plain from the construction that, for L/K a finite extension and $\underline{a} \in \mathbb{B}_K^0 \subset \mathbb{B}_L^0$, one has

$$J_L(\underline{a}) = J_K(\underline{a}) \circ \mathbb{N}_{L/K} \quad .$$

8.2.7 Stickelberger's theorem implies that the infinity type of $J_K(\underline{a})$ is $\theta_K(\underline{a})$ - defined in 8.2.1 - ,which takes values in $\mathbb{Z}$ if $\underline{a} \in \mathbb{B}^0$.

8.3 Example 8.1 revisited

8.3.1 Let us first write our basic characters of 8.1 in Anderson's notations. So let $K = \mathbb{Q}(\sqrt{-D})$ be of discriminant $-D<-8$. Put

$$\underline{a}_D = \sum_{\substack{j=1 \\ \varepsilon(j)=-1}}^{D} [\tfrac{j}{D}] \quad .$$

We find (8.1.3) that

$$n_1 = \langle \underline{a}_D \rangle \qquad \text{and} \qquad n_c = \langle c\underline{a}_D \rangle \quad .$$

Therefore, by the remark following 8.1.5, one has

$$\underline{a}_D \in \mathbb{B}^0 \quad .$$

Since $\underline{a}_D$ clearly belongs to $\mathbb{B}_K$, the character $J_K(\underline{a}_D)$ is well defined in Anderson's setup, and it is an easy exercise to check that

$$J_D = J_K(\underline{a}_D) \quad .$$

8.3.2 We shall now define, in Anderson's notation, a basic Jacobi sum character for each of the imaginary quadratic fields not treated in 8.1 and 8.3.1, i.e., for $D = 3,4,8$.

In all three cases the class number h_D is 1, and (in analogy with 8.1.5 for D = 3,4) we shall define J_D so that its infinity type T_D is $1.1 + 0.c = 1$. All we have to do is give an element $\underline{a}_D \in \mathbb{B}^0_{\mathbb{Q}(\sqrt{-D})}$, for D = 3,4,8, such that

$$\theta_{\mathbb{Q}(\sqrt{-D})}(\underline{a}_D)(\sigma) = \begin{cases} 1 & \text{if} \quad \sigma = 1 \\ 0 & \text{if} \quad \sigma = c \end{cases} .$$

For D = 3,4, we have tried to make a "classical" choice of $\underline{a}_D$ - see I. 7.5.
We propose as basic characters, $J_D = J_K(\underline{a}_D)$ with

$$\underline{a}_3 = 2[\tfrac{2}{3}] - [\tfrac{1}{3}]$$

$$\underline{a}_4 = [\tfrac{1}{2}] + [\tfrac{3}{4}] - [\tfrac{1}{4}]$$

$$\underline{a}_8 = -[\tfrac{1}{2}] + [\tfrac{5}{8}] + [\tfrac{7}{8}] .$$

8.4 The Stickelberger ideal

8.4.1 Definition. Let K be an abelian number field. The Stickelberger ideal of K is the ideal of the group ring $\mathbb{Z}[\mathrm{Gal}(K/\mathbb{Q})]$ consisting precisely of the infinity types of all Jacobi sum Hecke characters of K. It is denoted St_K .

It is not hard to check that our Stickelberger ideal St_K coincides with the one defined by Sinnott in [Sin] .
The main property of St_K which we shall have occasion to use is the following

8.4.2 Proposition. Let $A_K \subset \mathbb{Z}[\mathrm{Gal}(K/\mathbb{Q})]$ be the set of infinity types of all algebraic Hecke characters of K . Then St_K is a subgroup of finite index in A_K.

See [Sin], Theorem 2.1.

In general, it is very hard to describe St_K inside A_K, and even to give an explicit formula for the index $[A_K:St_K]$. An interesting case in which both can be done takes us back to our initial example of 8.1, resp. to 8.3.2.

8.4.3 Lemma *Let* $K = \mathbb{Q}(\sqrt{-D})$ *be any imaginary quadratic number field*, $-D$ *its discriminant. Then* St_K *consists precisely of the types*

$$k \cdot T_D + j \cdot (1 + c)$$

with $k, j \in \mathbf{Z}$. *The index* $[A_K:St_K] = h_D$.

(Recall that T_D was defined in 8.1.5, resp. 8.3.2.)

Proof. First observe that the given types are actually contained in St_K. This is true by construction for T_D, and $1 + c$ is the infinity type of the norm $\mathbb{N}$, i.e. of (say) $J_K(\underline{a}_3 + c\underline{a}_3)$, where $\underline{a}_3$ is as in 8.3.2 (but $K = \mathbb{Q}(\sqrt{-D})$ - cf. 8.2.6).

Secondly, the index of the set of types described is h_D. In fact, (8.1.5)

$$T_D - \frac{1}{2}\left(\frac{\varphi(D)}{2} - h_D\right) \cdot (1 + c) = h_D \cdot 1 ,$$

unless $D = 8$ - in which case $A_K = St_K$ according to 8.3.2. On the other hand, it follows from Theorem 2.1 combined with Theorem 5.3 of [Sin] that $[A_K:St_K] = h_D$, in our case. - But, to be sure, our quadratic fields do not really merit this quote: In fact, suppose $\underline{b} \in \mathbb{B}_K^0$, and $K = \mathbb{Q}(\sqrt{-D})$ with $D > 8$. Put $m = m(\underline{b})$ and decompose $\underline{b} = \sum_{1<d|m} \underline{b}_d$ with

$$\underline{b}_d = \sum_{\substack{i=1 \\ (i,m) = \frac{m}{d}}}^{m} n_i \left[\frac{i}{m}\right] .$$

Then $\underline{b}_d \in \mathbb{B}_{\mathbb{Q}(\mu_d)}$, and since the action of $\mathrm{Gal}(\overline{\mathbb{Q}}/\mathbb{Q})$ respects this decomposition of $\underline{b}$ we find that

$$\underline{b}_d \in \begin{cases} \mathbb{B}_{\mathbb{Q}} & \text{if} \quad K \not\subset \mathbb{Q}(\mu_d) \\ \mathbb{B}_K & \text{if} \quad K \subset \mathbb{Q}(\mu_d) \end{cases} .$$

In the first case, if $d \neq 2$, it follows that $\underline{b}_d$ is a multiple of $\sum_{\substack{j=1 \\ (j,d)=1}}^{d} [\frac{j}{d}]$, and therefore in $\mathbb{B}^0_{\mathbb{Q}}$. So it contributes to $J_K(\underline{b})$ a Hecke character of $\mathbb{Q}$, i.e., a multiple of $1 + c$ to the infinity type. We are therefore reduced to elements $\underline{b}$ of the form

$$\underline{b} = n_{\frac{m}{2}} [\tfrac{1}{2}] + \sum_{D|d|m} \underline{b}_d .$$

Now $\underline{b}_d \in \mathbb{B}_K$ implies that

$$\langle \underline{b}_d \rangle = r \cdot \sum_{\substack{j=1 \\ (j,d)=1 \\ \varepsilon(j)=-1}}^{d} \frac{j}{d} + s \cdot \sum_{\substack{j=1 \\ (j,d)=1 \\ \varepsilon(j)=+1}}^{d} \frac{j}{d} = r \cdot n_1(d) + s \cdot n_c(d),$$

with $n_1(D) = n_1$ and $n_c(D) = n_c$, as in 8.1.3.
Now it is easy to check that

$$n_1(d) - n_c(d) = (n_1 - n_c) \prod_{p|d} (1 - \varepsilon(p)) = h_D \cdot \prod_{p|d} (1-\varepsilon(p)) .$$

This means that, if $n \cdot 1$ is the infinity type of a Jacobi sum Hecke character of K , then $h_D | n$.

q.e.d.

CHAPTER ONE:
Motives for Algebraic Hecke Characters

This chapter contains an exposition of the less elementary and more geometric parts of the theory of algebraic Hecke characters. None of the results is original, but all the main theorems are fairly recent, so this is almost the first time that they are explicitly put together with a view to providing a "motivic" theory of Hecke characters. Compare however [A2] and [B1]. More precisely, we indicate a proof of Conjecture 8.1 of [DP] in the setting of certain motives for absolute Hodge cycles. We start out with the key example of the theory:

1. Abelian varieties with complex multiplication

1.1 Let K and E be two number fields (of finite degree over $\mathbb{Q}$), and let A be an abelian variety defined over K such that $2 \dim A = [E : \mathbb{Q}]$. Denote by $\mathrm{End}_K A$ the ring of endomorphisms of A that are defined over K, and assume there is an embedding of $\mathbb{Q}$-algebras

$$E \hookrightarrow \mathbb{Q} \otimes_{\mathbb{Z}} \mathrm{End}_K A,$$

<u>which will be fixed throughout</u>. Then we say that A/K has complex multiplication by E. For any prime power ℓ^n in $\mathbb{Z}$, denote by $A[\ell^n]$ the kernel of multiplication by ℓ^n on A, and define as usual

$$T_\ell(A) = \varprojlim_n A[\ell^n](\overline{K}), \qquad V_\ell(A) = T_\ell(A) \otimes_{\mathbb{Z}_\ell} \mathbb{Q}_\ell .$$

Here $\overline{K}$ is some fixed algebraic closure of K. There is a natural faithful action of $\mathrm{End}\, A$ on $T_\ell(A)$, and therefore of E on $V_\ell(A)$. As K is of characteristic 0, $T_\ell(A)$ is a free $\mathbb{Z}_\ell$ module of rank $2 \dim A$, and $V_\ell(A)$ is a free $E \otimes \mathbb{Q}_\ell$ module of rank 1. The action of $\mathrm{End}_K A$ commutes with the natural action of $\mathrm{Gal}(\overline{K}/K)$ on $T_\ell(A)$ and $V_\ell(A)$. So the Galois representation on $V_\ell(A)$ is $E \otimes \mathbb{Q}_\ell$-linear, and splits up as a sum of 1-dimensional λ-adic representations, for the places λ of E dividing ℓ,

$$\chi_\lambda : \mathrm{Gal}(\overline{K}/K) \longrightarrow E^*_\lambda .$$

<u>1.2</u> The formation of T_ℓ and V_ℓ is, of course, not restricted to abelian varieties with complex multiplication by a field E as above. And the part of the "Weil-conjectures" proved by Weil himself implies that the system of Galois representations $T_\ell(A)$, for ℓ varying over all rational primes, is <u>a strictly compatible system of</u> ($\mathbb{Q}$-) <u>rational representations</u>. Recall what this means: For a finite place $\mathfrak{p}$ of K, choose a decomposition group $D_\mathfrak{p} \subseteq \mathrm{Gal}(\overline{K}/K)$ and an element $\mathrm{Frob}\,\mathfrak{p} \in D_\mathfrak{p}$ whose class modulo the inertia subgroup $I_\mathfrak{p} \subseteq D_\mathfrak{p}$ corresponds to the automorphism $x \mapsto x^{1/\mathbb{N}\mathfrak{p}}$ of the residue field extension defined by $D_\mathfrak{p}$. Then $\mathrm{Frob}\,\mathfrak{p}$ is well defined modulo $I_\mathfrak{p}$, and up to conjugation in $\mathrm{Gal}(\overline{K}/K)$, and is called a <u>geometric Frobenius element at</u> $\mathfrak{p}$. With these notations the strict compatibility of the $T_\ell(A)$ means that there is a finite set S of places of K - to wit, the places where A has bad reduction - such that for all rational primes ℓ,ℓ' and any $\mathfrak{p}$ with $\mathfrak{p} \notin S$ and $\mathfrak{p} \nmid \ell\ell'$, $T_\ell(A)$ and $T_{\ell'}(A)$ are unramified at $\mathfrak{p}$, and the "characteristic polynomials" (which are then independent of the choice of $\mathrm{Frob}\,\mathfrak{p}$)

$$\det(1 - \mathrm{Frob}\,\mathfrak{p} \cdot X|T_\ell(A)) \quad \text{and} \quad \det(1 - \mathrm{Frob}\,\mathfrak{p} \cdot X|T_{\ell'}(A))$$

have coefficients in $\mathbb{Q}$ and are equal. - Cf. [ST]. Weil's theorems also tell us that all the eigenvalues of $\mathrm{Frob}\,\mathfrak{p}$ on $T_\ell(A)$ are algebraic numbers purely of absolute value $(\sqrt{\mathbb{N}\mathfrak{p}})^{-1}$.

<u>1.3</u> In the case of complex multiplication, <u>the system</u> (χ_λ) <u>of</u> λ-<u>adic representations</u>, λ <u>varying over all finite places of</u> E, <u>is itself a strictly compatible system of</u> E-<u>rational Galois representations</u>. That is to say, for every finite place $\mathfrak{p}$ of K not in the set of bad reduction S, there is a number $\chi(\mathfrak{p}) \in E^*$ such that for any finite place λ of E with $\mathfrak{p} \nmid \mathbb{N}\lambda$, $\chi(\mathfrak{p})$ maps to $\chi_\lambda(\mathrm{Frob}\,\mathfrak{p})$ under $E \hookrightarrow E_\lambda$. To prove this, one has to study the reduction mod $\mathfrak{p}$ of the abelian variety A: The Galois action of $\mathrm{Frob}\,\mathfrak{p}$ reduces to the

geometric Frobenius endomorphism on the reduction $\tilde{A}_{\mathfrak{p}}$ over the residue class field of K at $\mathfrak{p}$. This endomorphism lies in the center of the algebra $\mathbb{Q} \otimes_{\mathbb{Z}} \mathrm{End}(\tilde{A}_{\mathfrak{p}})$, and therefore lifts back to an element $\chi(\mathfrak{p}) \in E \subseteq \mathbb{Q} \otimes_{\mathbb{Z}} \mathrm{End}_K A$. $\chi(\mathfrak{p})$ is unique because reduction of endomorphisms is injective. It is the number sought. Cf. [LCM]. 2 §3.

It follows from the theory of Shimura and Taniyama - see [ShT], cf. [LCM], Chap. 4 - that χ extends multiplicatively to an algebraic Hecke character, i.e., "χ admits a conductor". But we prefer to deduce this from a much more general result which will be used later on:

<u>1.4 Proposition</u>: <u>Let</u> $\chi_\lambda : \mathrm{Gal}(\bar{K}/K) \longrightarrow E_\lambda^*$, <u>for all finite places</u> λ <u>of</u> E, <u>be a strictly compatible system of E-rational</u> λ<u>-adic representations of</u> K. <u>Then there is an algebraic Hecke character</u> χ <u>of</u> K <u>with values in</u> E <u>such that, for every finite place</u> λ <u>of</u> E, χ_λ <u>is the</u> λ<u>-adic representation attached to</u> χ (<u>defined in Chapter</u> 0, §5).

This proposition is just a variant of the main theorem of [Henn], which in turn is a corollary of a result in transcendence theory. In fact, Henniart proves that any abelian semisimple E-rational λ-adic Galois representation of K is locally algebraic. This means that there is a homomorphism of group-schemes over E_λ,

$$T_\lambda : Z/E_\lambda \longrightarrow \mathbb{G}_m/E_\lambda \quad (\text{where } Z = R_{K/\mathbb{Q}}\mathbb{G}_m)$$

such that the restriction of $T_\lambda/E_\lambda : Z(E_\lambda) \longrightarrow E_\lambda^*$ to the subgroup

$$\prod_{v|N\lambda} K_v^* = Z(\mathbb{Q}_\ell) \subseteq Z(E_\lambda)$$

coincides with the reciprocal of the composite map

$$\prod_{v|N\lambda} K_v^* \longrightarrow K_{\mathbb{A}}^* \xrightarrow{\text{Frob}} \mathrm{Gal}(K^{ab}/K) \xrightarrow{\text{repres.}} E_\lambda^*$$

on a suitable neighbourhood of 1. - Note that this condition is the analogue, for a finite place λ, of the existence of a defining ideal for the representation: see [Sℓ], III - 2. This is the reason why propositon 1.4 follows from Henniart's theorem.

<u>1.5</u> Let us come back to the abelian variety A over K with complex multiplication by E. Let χ be the algebraic Hecke character of K with values in E giving the λ-adic representations χ_λ of A, i.e., giving the action of $\mathrm{Gal}(\overline{K}/K)$ on the torsion points of A.

The Tate-conjecture proved by Faltings - cf. [Sch2], in particular 4.2 - implies that, for every ℓ, the $\mathbb{Q}_\ell$-subalgebra of $\mathrm{End}_{\mathbb{Q}_\ell} V_\ell(A)$ generated by the action of $\mathrm{Gal}(\overline{K}/K)$ is the commutant of $\mathrm{End}_K A \otimes_{\mathbb{Z}} \mathbb{Q}_\ell$. Since $E \otimes \mathbb{Q}_\ell$ is its own commutant in $\mathrm{End}_{\mathbb{Q}_\ell} V_\ell(A)$ it follows that $E = \mathbb{Q}(\chi)$ - the field generated over $\mathbb{Q}$ by the values of χ - if and only if $E = \mathbb{Q} \otimes_{\mathbb{Z}} \mathrm{End}_K A$, i.e., if and only if A is simple over K. In particular, E <u>is a</u> CM-<u>field in that case</u> (it cannot be totally real, as $\chi\overline{\chi} = \mathbb{N}^{-1}$) - but this proof is of course a little heavy handed for this elementary fact.

<u>1.6</u> The character χ has weight -1, and, moreover, its infinity-type T is what we call a CM-type of K (see 0 §3). To determine $T : K^* \longrightarrow E^*$, extend it to a map from ideals of K to ideals of E so that

$$\chi(\mathfrak{p}) \cdot \mathcal{O}_E = T(\mathfrak{p}),$$

for almost all prime ideals $\mathfrak{p}$ of K. The prime ideal decomposition of $\chi(\mathfrak{p}) \cdot \mathcal{O}_E$ can be determined from the fact that

$$\chi(\mathfrak{p}) \in E \subseteq \mathbb{Q} \otimes_{\mathbb{Z}} \mathrm{End}_K A \longrightarrow \mathbb{Q} \otimes_{\mathbb{Z}} \mathrm{End}(\tilde{A}_{\mathfrak{p}})$$

reduces to the geometric Frobenius on $\tilde{A}_{\mathfrak{p}}$, by letting $\mathrm{End}\,A$ and $\mathrm{End}\,\tilde{A}_{\mathfrak{p}}$ act on the tangent spaces $\mathrm{Lie}\,A$ and $\mathrm{Lie}\,\tilde{A}_{\mathfrak{p}}$ - see [Gi]. Viewing $\mathrm{Lie}\,A$ as a $K \otimes_{\mathbb{Q}} E$ module - cf. [ST], §7 - the final result can be stated like this;

$$T(x) = \det_E(x \otimes 1 \mid \text{Lie } A)^{-1} \in E^*.$$

Recall in passing that the algebraic homomorphism

$$E^* \longrightarrow K^*$$
$$y \longmapsto \det_K(1 \otimes y \mid \text{Lie } A)^{-1}$$

(or rather, its reciprocal) is often called the CM-type of A, and T (or rather, -T) is called its "reflex-type" (on K).

1.7 An interesting way of rephrasing this description of the infinity-type of χ is provided by the Hodge decomposition of the first singular homology of A. For $\sigma : K \longrightarrow \mathbb{C}$, write

$$H_1^\sigma(A,\mathbb{C}) = H_1((A \times_{K,\sigma} \mathbb{C})(\mathbb{C}),\mathbb{Z}) \otimes_{\mathbb{Z}} \mathbb{C} = H_\sigma^{-1,0} \oplus H_\sigma^{0,-1}.$$

Then

$$H_\sigma^{0,-1} = \text{Lie}(A \times_{K,\sigma} \mathbb{C}) = (\text{Lie } A) \otimes_{K,\sigma} \mathbb{C}.$$

But Lie A is also an E module. For any $\tau : E \longrightarrow \mathbb{C}$, define $n(\sigma,\tau)$ to be -1 or 0 according as the action of E on Lie A agrees with the action of E via $E \xrightarrow{\tau} \mathbb{C}$ on the subspace $H_\sigma^{-1,0}$, or not (in which case $n(\sigma,\bar{\tau})$ will be -1). These integers $n(\sigma,\tau)$ describe the Hodge decompositions of the $H_1^\sigma(A)$, for all σ, as follows: Since $E \subseteq \mathbb{Q} \otimes_{\mathbb{Z}} \text{End}_K A$, every $H_1^\sigma(A,\mathbb{Q})$ has the structure of an E vector space (of dimension 1). The direct factor $H_1^\sigma(A,\mathbb{Q}) \otimes_{E,\tau} \mathbb{C}$ of $H_1^\sigma(A,\mathbb{C})$ lies in $H_\sigma^{n(\sigma,\tau),(-1-n(\sigma,\tau))}$.

On the other hand, the identity $T(x) = \det_E(x \otimes 1 \mid \text{Lie } A)^{-1}$ means that the $n(\sigma,\tau)$'s are precisely the integers attached to χ in Chapter 0, §4. Later on in this chapter, we shall systematically generalize this kind of correspondence between infinity-types and Hodge structures on an E-vector space of dimension one.

1.8 Sticking to our preference for the geometric Frobenius

over the arithmetic one, and taking the $V_\ell(A)$ (rather than their duals $H^1_{ét}(A/\bar{K},\mathbb{Q}_\ell)$) as the ℓ-adic representations characterizing A, we are led, by the motivic formalism explained in 6.5.9 below, to define the (false!) "Hasse-Weil" L-function of A over K, for $\mathrm{Re}(s) > \frac{1}{2}$, by:

$$L(A/K,s) = \prod_{\mathfrak{p}} \det(1 - \mathrm{Frob}\,\mathfrak{p} \cdot \mathbb{N}\mathfrak{p}^{-s} | V_\ell(A)^{I_\mathfrak{p}})^{-1},$$

where $I_\mathfrak{p}$ is an inertia-subgroup at $\mathfrak{p}$; $\mathfrak{p}$ runs over all finite places of K, and it is understood that the determinant is calculated using some prime number ℓ such that $\mathfrak{p} \nmid \ell$. - Cf. [ST].

Since $V_\ell(A) = \bigoplus_{\lambda|\ell} X_\lambda$, we see immediately that

$$L(A/K,s) = \prod_{\tau:E\to\mathbb{C}} L(\chi^\tau,s),$$

in the notation of 0 §6.

<u>1.8.1</u> <u>Warning</u>: Our function $L(A/K,s)$ is <u>not</u> the usual Hasse-Weil L-function of A over K as employed for instance in the standard formulation of the conjectures of Birch and Swinnerton-Dyer:

$$L_{\mathrm{usual}}(A/K,s) = \prod_{\mathfrak{p}} \det(1 - \mathrm{Frob}\,\mathfrak{p} \cdot \mathbb{N}\mathfrak{p}^{-s} | H^1_{ét}(A/\bar{K},\mathbb{Q}_\ell)^{I_\mathfrak{p}})^{-1}$$

$$= \prod_{\mathfrak{p}} \det(1 - \mathrm{Frob}_{\mathrm{arith}}\mathfrak{p} \cdot \mathbb{N}\mathfrak{p}^{-s} | V_\ell(A)_{I_\mathfrak{p}})^{-1},$$

with $V_\ell(A)_{I_\mathfrak{p}}$ denoting coinvariants under inertia. Under the functional equation (which for general A is only conjectured), $L(A/K,s)$ corresponds to $L_{\mathrm{usual}}(A/K,1-s)$.

The L-function $L(A/K,s)$ is attached to A/K without reference to the fact that A has complex multiplication by E. In the presence of complex multiplication it is, however, more adequate to consider the array of L-functions

$$L^*(A/K,s) = L^*(\chi,s) = (L(\chi^\tau,s))_{\tau:E\to\mathbb{C}}$$

taking values in $E \otimes_{\mathbb{Q}} \mathbb{C}$.

In order to find "geometric" objects over K whose L-functions include all L-functions of algebraic Hecke characters of K, we have to pass from abelian varieties (with complex multiplication) to motives.

2. Motives for absolute Hodge cycles

The lecture notes [DMOS] contain the first detailed exposition of a theory of motives over fields of characteristic zero which does not depend on unproven conjectures. They will be our constant frame of reference when we are dealing with motives. The other main source for the kind of questions treated here is of course Deligne's article [DP] which, however, insists on the general formalism, not attaching any specific meaning to the word motive, and using a hierarchy of conjectures when needed. In this section, we shall quickly review the main concepts and results from the general theory of motives as constructed in [DMOS], II § 6.- For a somewhat different setup of largely the same theory, see [A2].

2.1 ABSOLUTE HODGE CYCLES (Reference: [DMOS],I § 1, § 2)

Let K be a field which can be embedded into $\mathbb{C}$, and X a smooth projective algebraic variety over K. To every place of $\mathbb{Q}$, we can attach a cohomology theory of varieties X over K:

At infinity, take the algebraic de Rham cohomology

$$H^{\cdot}_{DR}(X) = H^{\cdot}_{DR}(X/K) = H^{\cdot}(X_{Zar}, \Omega^{\cdot}_{X/K}).$$

For all n, $H^n_{DR}(X)$ is a K-vector space equipped with a descending filtration $F^{\cdot}$, the Hodge filtration.

All the finite primes ℓ of $\mathbb{Q}$ can be treated simultaneously: denote by $\mathbb{Q}_{\mathbb{A}^f}$ the ring of finite adèles of $\mathbb{Q}$, and put

$$H^{\bullet}_{\mathbf{A}f}(X) = H^{\bullet}_{\mathbf{A}f}(X/K) = \{\varprojlim_m H^{\bullet}((X\times_K \overline{K})_{\text{ét}}, \mathbf{Z}/m\mathbf{Z})\} \otimes_{\hat{\mathbf{Z}}} \mathbb{Q}_{\mathbf{A}f} ,$$

where $\overline{K}$ is an algebraic closure of K. The $H^n_{\mathbf{A}f}(X/K)$ are $\mathbf{Q}_{\mathbf{A}f}$-modules with a natural action of $\mathrm{Gal}(\overline{K}/K)$. The $\mathbf{Q}_\ell$-component of $H^{\bullet}_{\mathbf{A}f}(X)$ will be written $H^{\bullet}_{\ell}(X)$.

For any embedding $\sigma: K \to \mathbb{C}$, denote by σX the extension of scalars $X\times_{K,\sigma}\mathbb{C}$, and by

$$H^{\bullet}_{\sigma}(X) = H^{\bullet}(\sigma X(\mathbb{C}), \mathbf{Q})$$

the rational singular cohomology (resp., $H^{\sigma}_{\bullet}(X)$ the rational singular homology) of $\sigma X(\mathbb{C})$. The $H^{\bullet}_{\sigma}(X)$ are rational Hodge structures, i.e., $\mathbb{Q}$-vector spaces together with a decomposition of the complexifications

$$H^n_{\sigma}(X) \otimes_{\mathbb{Q}} \mathbb{C} = \bigoplus_{p+q=n} H^{pq}$$

such that H^{pq} and H^{qp} are interchanged by complex conjugation. Whenever K is given as a subfield of $\mathbb{C}$ (e.g., $K = \mathbb{Q}, \mathbb{R}$), $H^{\bullet}_{\sigma}$, for σ the inclusion $K \subset \mathbb{C}$, will be written $H^{\bullet}_B$, the letter B standing for "Betti". (See also 6.0.)

In these cohomology theories, define the Tate twist as follows (we write $\mu_m(\overline{K}) = \{\zeta \in \overline{K}^* \mid \zeta^m = 1\}$).

$\mathbb{Q}_{DR}(1) = K$	$\mathbb{Q}_{DR}(-1) = K$
$\mathbb{Q}_{DR}(1) = F^{-1} \supset F^0 = 0$	$\mathbb{Q}_{DR}(-1) = F^1 \supset F^2 = 0$
$\mathbb{Q}_{\mathbf{A}f}(1) = \varprojlim_m \mu_m(\overline{K}) \otimes_{\mathbf{Z}} \mathbb{Q}_{\mathbf{A}f}$	$\mathbb{Q}_{\mathbf{A}f}(-1) = \mathrm{Hom}(\mathbb{Q}_{\mathbf{A}f}(1), \mathbb{Q}_{\mathbf{A}f})$
$\mathbb{Q}_B(1) = 2\pi i\, \mathbb{Q} \subset \mathbb{C}$	$\mathbb{Q}_B(-1) = \frac{1}{2\pi i} \cdot \mathbb{Q}$
$\mathbb{Q}_B(1) \otimes_{\mathbb{Q}} \mathbb{C} = H^{-1,-1}$	$\mathbb{Q}_B(-1) \otimes \mathbb{C} = H^{1,1}$
the involution F_∞ (see II, 1.6.1) acts on $\mathbb{Q}_B(1)$ as -1	the involution F_∞ acts on $\mathbb{Q}_B(-1)$ as -1
For $m \in \mathbf{Z}, m > 0$,	For $m \in \mathbf{Z}, m < 0$,

$\mathbb{Q}_{\cdot\cdot}(m) = \mathbb{Q}_{\cdot\cdot}(1)^{\otimes m}$ $\qquad\qquad$ $\mathbb{Q}_{\cdot\cdot}(m) = \mathbb{Q}_{\cdot\cdot}(-1)^{\otimes -m}$

For all $m \in \mathbb{Z}$,

$$H^{\cdot}_{\cdot\cdot}(X)(m) = H^{\cdot}_{\cdot\cdot}(X) \otimes_{\cdot\cdot} \mathbb{Q}_{\cdot\cdot}(m).$$

For every embedding $\sigma:\overline{K} \hookrightarrow \mathbb{C}$ of the fixed algebraic closure $\overline{K}$ of K into $\mathbb{C}$, there is the total <u>comparison isomorphism</u>

(2.1.1) $\quad H^{\cdot}_{\sigma}(X)(m) \otimes_{\mathbb{Q}} (\mathbb{C} \times \mathbb{A}^{f}) \overset{\cong}{\to} H^{\cdot}_{DR}(\sigma X) \times H^{\cdot}_{\mathbb{A}^f}(\sigma X)(m)$,

the filtration on $H^{\cdot}_{DR}(\sigma X) = H^{\cdot}_{DR}(X) \otimes_{K,\sigma} \mathbb{C}$ being given by

$$\bigoplus_{p' \geq p} H^{p',q'} \overset{\cong}{\to} F^{p}.$$

Note that σ induces an isomorphism

$$H^{\cdot}_{\mathbb{A}^f}(X \times_K \overline{K}/K) \cong H^{\cdot}_{\mathbb{A}^f}(\sigma X/\mathbb{C}),$$

and $H^{\cdot}_{\mathbb{A}^f}(X \times_K \overline{K}/\overline{K})$ is just $H^{\cdot}_{\mathbb{A}^f}(X/K)$, with the action of $Gal(\overline{K}/K)$ forgotten.

Abbreviate $H^{\cdot}_{DR}(X)(m) \times H^{\cdot}_{\mathbb{A}^f}(X)(m)$ to $H^{\cdot}_{\mathbb{A}}(X)(m)$. For $p \in \mathbb{Z}$, $p > 0$, an element $t \in H^{2p}_{\mathbb{A}}(X/K)(p)$ is called a <u>Hodge cycle</u> (<u>of codimension</u> p) <u>over</u> K <u>relative to</u> $\sigma:\overline{K} \to \mathbb{C}$, if

(i) $\quad t \in H^{2p}_{\sigma}(X)(p) \subset H^{2p}_{\sigma}(X)(p) \otimes (\mathbb{C} \times \mathbb{Q}_{\mathbb{A}^f})$ (by 2.1.1),

(ii) the H_{DR}-component t_{DR} of t lies in

$$F^{0}H^{2p}_{DR}(X)(p) = F^{p}H^{p}_{DR}(X).$$

The algebraic condition (ii) is clearly equivalent to the analytic one:

(ii)' $\quad t_{DR} \in H^{0,0} \subset H^{2p}_{\sigma}(X)(p) \otimes \mathbb{C}$ (by $2.1.1_{DR}$).

Condition (i) means that the components of $t \in H^{2p}_{\mathbb{A}}(X)(p)$ all correspond to a single element in $H^{2p}_{\sigma}(X)(p)$, under the various comparison isomorphisms: between Betti and de Rham, Betti and étale cohomologies.

An absolute Hodge cycle on X over K (of codimension p) is an element $t \in H^{2p}_{\mathbb{A}}(X/K)(p)$ which is a Hodge cycle relative to all $\sigma:\overline{K} \to \mathbb{C}$. The $\mathbb{Q}$-vector space of all these absolute Hodge cycles is denoted by $C^p_{AH}(X/K)$, or $C^p_{AH}(X)$ if the reference to K is clear or irrelevant (e.g., $K = \overline{K}$ - see below). Clearly, $\dim_{\mathbb{Q}} C^p_{AH}(X/K) < \infty$. The definition of C_{AH} we have given does not easily betray its virtues. - Looked at "from the side of Betti cohomology", $C^p_{AH}(X/K)$ is isomorphic to the $\mathbb{Q}$-vector space of arrays $(t_\sigma | \sigma:\overline{K} \to \mathbb{C})$, where

(i) $\quad t_\sigma \in H^{2p}_\sigma(X)(p) \cap H^{0,0}$,

such that, fixing any $\sigma:\overline{K} \to \mathbb{C}$, we have for all $s \in \operatorname{Aut} \mathbb{C}$:

(ii) $\quad t_\sigma$ and $t_{s\circ\sigma}$ correspond to the same element $t_{DR} \in H^{2p}_{DR}(X/\overline{K})$ under the Betti-de Rham comparison isomorphism

$$\begin{array}{ccc} t_\sigma \in H^{2p}_{DR}(X/\overline{K}) \otimes_{\overline{K},\sigma} \mathbb{C} & \longleftarrow & \\ \big\downarrow (\mathrm{id}\otimes s) & & H^{2p}_{DR}(X/\overline{K})(p) \ni t_{DR} \\ t_{s\circ\sigma} \in H^{2p}_{DR}(X/\overline{K}) \otimes_{\overline{K},s\circ\sigma} \mathbb{C} & \longleftarrow & \end{array}$$

(iii) $\quad t_\sigma$ and $t_{s\circ\sigma}$ correspond to the same element $t_{\mathbb{A}^f} \in H^{2p}_{\mathbb{A}^f}(X)(p)$ under the Betti-étale comparison isomorphism:

$$\begin{array}{ccc} t_\sigma \in H^{2p}_{\mathbb{A}^f}(\sigma X)(p) & \overset{\cong}{\longleftarrow} & \\ \big\downarrow s & & H^{2p}_{\mathbb{A}^f}(X/\overline{K})(p) \ni t_{\mathbb{A}^f} \\ t_{s\circ\sigma} \in H^{2p}_{\mathbb{A}^f}((s\circ\sigma)X)(p) & \overset{\cong}{\longleftarrow} & \end{array}$$

(iv) $\quad t_\sigma = t_{\sigma\circ g}$, for all $g \in \operatorname{Gal}(\overline{K}/K)$.

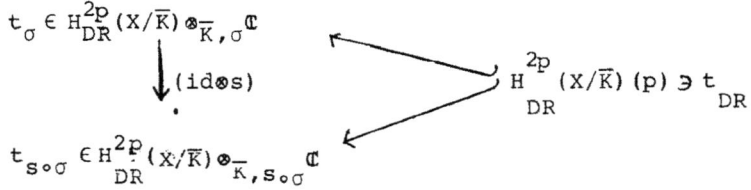

Note that (iv) makes sense because X is defined over K, and so $H^{\bullet}_{\sigma}(X) = H^{\bullet}_{\sigma\circ g}(X)$. Given the compatibilities (ii) and (iii), the Galois-action may also be read on H_{DR} or $H_{\mathbb{A}^f}$, and (iv) may be replaced by either

(iv)' $t_{DR} \in H^{2p}_{DR}(X/K)$,

- which is what we used in our first definition of absolute Hodge cycles - , or by

(iv)'' $t_{\mathbb{A}^f} \in [H^{2p}_{\mathbb{A}^f}(X)(p)]^{Gal(\bar{K}/K)}$.

The following proposition sums up the fundamental rationality properties of absolute Hodge cycles.

2.1.2 Proposition: a) If $L \supset K$ is still embeddable into $\mathbb{C}$, then the natural map

$$C^p_{AH}(X/\bar{K}) \to C^p_{AH}(X/\bar{L})$$

is an ismorphism.

b) $Gal(\bar{K}/K)$ acts on $C^p_{AH}(X/\bar{K})$ through a finite quotient.

To prove a), one has to invoke the theory of the Gauss-Manin connection. As for b), we have already seen that $Gal(\bar{K}/K)$ stabilizes $C^p_{AH}(X/\bar{K})$. This action being continuous and $\mathbb{Q}$-linear on a finite dimensional $\mathbb{Q}$-vector space it factors through a finite quotient.

The crucial result justifying in a way the theory we are about to develop is Deligne's

2.1.3 Theorem: If K is algebraically closed, and X is an abelian variety over K, then every cycle $t \in H^{2p}_{\mathbb{A}}(X)(p)$ which is a Hodge cycle relative to one embedding $\sigma: K \to \mathbb{C}$ is an absolute Hodge cycle.

If K is not algebraically closed the conclusion will hold for cycles t whose H_{DR}- or $H_{\mathbb{A}^f}$-component is fixed by $Gal(\bar{K}/K)$.

The proof starts with the "exceptional" Hodge cycles on abelian varieties with some complex multiplications, studied in [Weil; 1977 c] and used more generally in [Gr 2]. They are shown to be absolute Hodge cycles by a deformation argument very much reminiscent of Gross' paper. From there, Deligne goes on to CM-abelian varieties first, and passes to the general case by another deformation argument. --See [DMOS], chap. I.

2.1.4 Remark. Every algebraic cycle, i.e., every element of $H^{2p}_{\mathbb{A}}(X)(p)$ coming from an algebraic subvariety of X of codimension p, via the cycle maps in de Rham and étale cohomology, is an absolute Hodge cycle. The Hodge conjecture states that any cycle which is a Hodge cycle relative to one σ is algebraic. In this sense Deligne's result proves part of the Hodge conjecture for abelian varieties.

2.2 MOTIVES (Reference: [DMOS],II § 6)

Let K as before be a field embeddable into $\mathbb{C}$. The construction of the category $\mathcal{M}_K$ of motives over K, via absolute Hodge cycles, proceeds roughly as follows.

Step 1. Let $\mathcal{CV}_K$ be the category with objects written $h(X)$, for X varying over smooth projective algebraic varieties defined over K, and morphisms the $\mathbb{Q}$-vector spaces defined by $\mathrm{Hom}(h(X),h(Y)) = C^n_{AH}(X\times Y)$, if X is connected of dimension n, and by additivity, via $h(X\amalg Y) = h(X)\oplus h(Y)$, in general.
To understand this definition of morphisms, note that

$$C^{n+p}_{AH}(X\times Y) \hookrightarrow H^{2(n+p)}_{\cdot\cdot}(X\times Y)(n+p) = \bigoplus_r \mathrm{Hom}_{\cdot\cdot}(H^r(X),H^{r+2p}(Y)(p))$$

(by Künneth and duality), so that $\mathrm{Hom}(h(X),h(Y))$ really gives a family of maps between the graded Betti, resp. de Rham, resp. étale cohomology of X and Y.

Taking the cycle of the graph of a K-morphism $X \to Y$ yields a contravariant functor $\mathcal{V}_K \to \mathcal{CV}_K$, where $\mathcal{V}_K$ is the usual category of smooth projective K-varieties.

It is essential to consider $\mathcal{CV}_K$ as a tensor category (cf. [DMOS], II § 1), the tensor product being given by

$$h(X) \otimes h(Y) = h(X \times Y),$$

with obvious associativity and commutativity constraints, and h(pt) as identity object.

Step 2. Let $\dot{\mathcal{M}}_K^+$ be the pseudoabelian (or: "Karoubian") envelope of $\mathcal{CV}_K$. This means we formally adjoint objects to $\mathcal{CV}_K$ to insure that every idempotent in End(h(X)), for any X, arises from a splitting $h(X) = M' \oplus M''$ in $\dot{\mathcal{M}}_K^+$. The objects of $\dot{\mathcal{M}}_K^+$ can be represented explicitly as pairs (M,p), with M in $\mathcal{CV}_K$ and $p \in \mathrm{End}(M)$, $p^2 = \mathrm{id}_M$. The morphisms are given by

$$\mathrm{Hom}((M,p),(N,q)) = \frac{\{f:M \to N \mid f \circ p = q \circ f\}}{\{f \mid f \circ p = 0 = q \circ f\}} .$$

For every X, there is a standard decomposition, in End(h(X)), of $\mathrm{id}_{h(X)}$ into a sum of pairwise orthogonal idempotents

$$\mathrm{id}_{h(X)} = p^0 + p^1 + p^2 + \ldots$$

(actually a finite sum): take p^r to be the projection

$$\oplus H^{\cdot}(X) \to H^r(X) ,$$

in all cohomology theories. In terms of absolute Hodge cycles, look at the Künneth components of the diagonal $\Delta \subset X \times X$:

$$H^{2n}(X \times X)(n) = \bigoplus_{i=0}^{2n} H^{2n-i}(X) \otimes H^i(X)$$

$$c\ell(\Delta) = \sum_{i=0}^{2n} \pi^i .$$

For all i, one has $\pi^i \in C_{AH}^n(X \times X)$.

So, for every $X \in \mathcal{V}_K$ and $0 \leq r \leq 2 \dim X$, there is an object $h^r(X) \in \dot{\mathcal{M}}_K^+$ which singles out the r-th cohomology groups of X. Whence a grading on the objects of $\dot{\mathcal{M}}_K^+$.

The tensor structure on $\dot{M}_K^+$ is defined by

$$(M,p) \otimes (N,q) = (M \otimes N, p \otimes q).$$

It respects the grading in the sense that the "Künneth formula" holds:

$$(X \otimes Y)^r = \bigoplus_{s+t=r} X^s \otimes Y^t ,$$

i.e., one has to check that the Künneth components are absolute Hodge cycles.

Step 3. We now introduce the Tate twist into $\dot{M}_K^+$. The motive $h^2(\mathbb{P}^1) \in \dot{M}_K^+$ has precisely the cohomology groups denoted by $\mathbb{Q}_{..}(-1)$ in 2.1. Let $\dot{M}_K$ be the category obtained from $\dot{M}_K^+$ by inverting the fully faithful functor $M \mapsto M \otimes h^2(\mathbb{P}^1)$ on $\dot{M}_K^+$. In down-to-earth terms, this means that the objects of M_K can be represented as pairs

$$(M,m) , \text{ with } M \in \dot{M}_K^+ \text{ and } m \in \mathbb{Z}.$$

For morphisms we have

$$\mathrm{Hom}((M,m),(N,n)) = \mathrm{Hom}_{\dot{M}_K^+}\big(M \otimes (h^2(\mathbb{P}^1))^{\otimes k-m}, N \otimes (h^2(\mathbb{P}^1))^{\otimes k-n}\big),$$

for any $k \geq m,n$.

This definition is independent of k, and thus allows to define the composition of morphisms by choosing k sufficiently large.

We write (M,m) as $M(m)$ or $M \otimes \mathbb{Q}(m)$. $\dot{M}_K^+$ is a full subcategory of $\dot{M}_K$ via $M \mapsto M(0)$. The tensor structure on $\dot{M}_K$ is given by

$$M(m) \otimes N(n) = (M \otimes N)(m+n) .$$

The grading on $\dot{M}_K^+$ extends to $\dot{M}_K$ by

$$M(m)^r = M^{r-2m} .$$

<u>Step 4.</u> $\dot{M}_K$ is almost the category of motives we want. Its only technical (but important) shortcoming is the sign convention relating the grading of the objects to the tensor structure. The point is that a good category of motives should be equivalent to the category of representations of a group scheme - i.e., should be a <u>tannakian category</u> (see 2.3 below!). In such a category, the rank of a representation (i.e., the trace of its identity-morphism - see [DMOS],II § 1.7) is simply the dimension of the underlying space, i.e., a positive integer. But in $\dot{M}_K$, the rank of $h(X)$ turns out to be the Euler-Poincaré characteristic which may of course be negative. - To put it another way, the problem is that the cup product which yields the identification of $h(X\times Y)$ with $h(X) \otimes h(Y)$ is not commutative.
This problem can be overcome by tampering with the commutativity constraint of the tensor structure on $\dot{M}_K$:

$$\dot{\Psi}: M \otimes N \tilde{\rightarrow} N \otimes M, \quad \dot{\Psi} = \oplus\, \dot{\Psi}^{p,q} \quad \text{where}$$

$$\dot{\Psi}^{p,q}: M^p \otimes N^q \tilde{\rightarrow} N^q \otimes M^p.$$

The corrected constraint is defined by:

$$\Psi: M \otimes N \tilde{\rightarrow} N \otimes M, \quad \Psi = \oplus \Psi^{p,q} \quad \text{where}$$

$$\Psi^{p,q} = (-1)^{pq}\ \dot{\Psi}^{p,q} .$$

$\dot{M}_K$ with the commutativity constraint $\dot{\Psi}$ replaced by Ψ , is denoted by M_K ,and called <u>the category of motives over</u> K (constructed with absolute Hodge cycles). M_K is a tannakian category, in the sense to be explained below.

<u>2.2.1.</u> One shows that M_K is a <u>semisimple</u> tannakian category (i.e., every exact sequence in M_K splits), see [DMOS], II 6.5, and that End M, for every object M of M_K is a semisimple $\mathbb{Q}$-algebra.

<u>2.2.2</u> For practical purposes, it is often sufficient to identify a motive M in M_K with the string of its realizations in the different cohomology theories:

$$H_\sigma(M), H_{DR}(M), H_\ell(M) \quad (\sigma: K \to \mathbb{C}; \ell \text{ a rational prime}).$$

These realizations are formally defined by extending the cohomology functors $H_{..}: \mathcal{V}_K \to A_{..}$ to $\mathcal{M}_K$, where $A_{..}$ is the corresponding target category: A_B= rational Hodge structures; A_{DR}= filtered finite-dimensional K-vector spaces; A_ℓ= finite dimensional $\mathbb{Q}_\ell$-vector spaces with $\mathrm{Gal}(\bar{K}/K)$-action. For each of the cohomology theories, the extension of $H_{..}$ is possible because the categories $A_{..}$ are Karoubian, have a tensor structure with Künneth formula, and the Tate twist is defined (see 2.1).
Then, Hom (M,N), for $M,N \in \mathcal{M}_K$, consists precisely of the systems of maps

$$(f_{\mathbb{A}}) = (f_{DR}, f_\ell \mid \text{all } \ell)$$

such that

$$f_{DR}: H_{DR}(M) \to H_{DR}(N)$$

is a K-linear map preserving the Hodge filtrations, and for every prime number ℓ,

$$f_\ell: H_\ell(M) \to H_\ell(N)$$

is a $\mathbb{Q}$-linear map with $f_\ell^\sigma = f_\ell$, for all $\sigma \in \mathrm{Gal}(\bar{K}/K)$, and such that, for any embedding $\sigma: K \to \mathbb{C}$, there exists a $\mathbb{Q}$-linear map

$$f_\sigma: H_\sigma(M) \to H_\sigma(N)$$

such that $f_\sigma \otimes (\mathbb{C} \times \mathbb{Q}_{\mathbb{A}f})$ corresponds to $f_{\mathbb{A}}$ under the comparison isomorphism (2.1.1)

$$H_\sigma(M) \otimes (\mathbb{C} \times \mathbb{Q}_{\mathbb{A}f}) \xrightarrow{\sim} H_{\mathbb{A}}(\sigma M).$$

(Here we have used the fact that the functor $\times_{K,\sigma}\mathbb{C}$, or in general $\times_K K': \mathcal{V}_K \to \mathcal{V}_{K'}$, extends to a functor $\mathcal{M}_K \to \mathcal{M}_{K'}$). Note that f_σ has to respect the Hodge decomposition.

f is an isomorphism, if at least one of f_σ, f_{DR}, f_ℓ is.

2.2.3 Remark: For any field extension $K' \supset K$, the functor extension of scalars, $\times_K K'$, carries over from varieties to motives, $\times_K K' : \mathcal{M}_K \to \mathcal{M}_{K'}$. If $K' \supset K$ is finite, so does the functor restriction of scalars $R_{K'/K} : \mathcal{V}_{K'} \to \mathcal{V}_K$, defined by

$$\begin{array}{ccccc} X & & X & & \\ \downarrow & \longmapsto & \downarrow & \searrow & \\ & & & & \operatorname{Spec} K' \\ \operatorname{Spec} K' & & \operatorname{Spec} K & \swarrow & \end{array}$$

(or classically, according to [We 4],1.3.) Note, however, that $R_{K'/K} : \mathcal{M}_{K'} \to \mathcal{M}_K$ is not $\otimes$-compatible.

2.3 TANNAKIAN PHILOSOPHY (Reference: [DMOS],II §§ 1-5; [Sa])

2.3.1 Let k be a field, and G an affine group scheme over k, i.e., a representable group valued functor on k-algebras, or again, the inverse limit of affine k-algebraic groups (=affine group schemes of finite type over k). - Cf. [Wa]. The category $\mathcal{R}ep_k(G)$ of finite dimensional representations of G over k(=algebraic morphisms $G \to GL(V)$, with a finite dimensional k-vector space V) has the following properties:

- $\mathcal{R}ep_k(G)$ is a k-linear abelian category;

- $\mathcal{R}ep_k(G)$ is a $\otimes$-category - cf. [DMOS], II § 1 - with commutativity and associativity constraint, unit object whose algebra of endomorphisms is k, $\otimes$-compatible <u>Hom</u> - objects, and duals such that each object is isomorphic to its double dual;

- there is a k-linear, $\otimes$-compatible functor

$$\omega : \mathcal{R}ep_k(G) \to \mathcal{V}ec_k \; ;$$

  ω is faithful, additive and exact.

Namely, take for ω the functor forgetting the G-action on V.

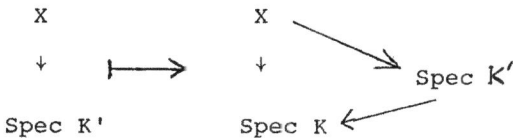

A <u>(neutralized)</u> <u>tannakian</u> <u>category</u> <u>(over</u> k) is a pair $(\mathcal{C},\omega)$ consisting of a $\otimes$-category $\mathcal{C}$ satisfying the first two properties listed for $\mathcal{R}ep_k(G)$ above, and a $\otimes$-functor $\omega:\mathcal{C} \to \mathcal{V}ec_k$ verifying the third set of conditions.

"Tannakian philosophy" exploits the fundamental theorem to the effect that there is a 1-1 correspondence between affine k-group schemes G and neutralized tannakian categories over k:

$$G \quad \to \quad (\mathcal{R}ep_k(G), \omega_{forget})$$

$$Aut^{\otimes}\omega \quad \leftarrow \quad (\mathcal{C},\omega),$$

where $Aut^{\otimes}\omega$ is the group valued functor on k-algebras R such that $(Aut^{\otimes}\omega)(R)$ consists of all R-linear, $\otimes$-compatible automorphisms of the functor $X \mapsto \omega(X)\otimes_k R$ on $\mathcal{C}$. - Thus, it is shown that $Aut^{\otimes}\omega$ can be represented by an affine group scheme G over k, and that ω defines an equivalence of $\otimes$-categories $\mathcal{C} \to \mathcal{R}ep_k(G)$.

<u>2.3.2</u> Here is a portion of the dictionary between affine group schemes and tannakian categories which results from the fundamental correspondence between them. - Cf. [Sa],p. 156 f; [DMOS], pp. 138-144; and [A2], 3.4 - 3.6.

Suppose $(\mathcal{C},\omega)$ and $(\mathcal{C}',\omega')$ are neutralized tannakian categories over k with corresponding affine k-group schemes G and G'. Any additive $\otimes$-compatible functor $F:\mathcal{C}' \to \mathcal{C}$ such that $\omega' = \omega \circ F$ induces a k-morphism $F^{\#}:G \to G'$.

(a) Suppose k is of characteristic 0. Then $\mathcal{C}$ is semisimple (i.e., every exact sequence in $\mathcal{C}$ splits) if and only if G, i.e., its connected component G°, is (pro-)reductive.

(b) Suppose the equivalent conditions of (a) are verified. Then $F^{\#}$ is faithfully flat if and only if F is fully faithful.

(c) $F^{\#}$ is a closed immersion if and only if every object of $\mathcal{C}$

is isomorphic to a subquotient of $F(X)$, for some object X of $\mathcal{C}'$.

(d) The objects $\{X_i \mid i \in I\}$ of $\mathcal{C}$ generate the tannakian category $\mathcal{C}$ (i.e., every object of $\mathcal{C}$ is isomorphic to an object obtained from the $\{X_i\}$ by a finite number of operations of the following kind: tensor product, dual, direct sum, subquotient) if and only if, for every k-algebra R, the obvious map

$$G(R) \to \prod_{i \in I} \mathrm{Aut}_R(\omega(X_i) \otimes_k R)$$

is injective.

<u>2.3.3</u> As an example of a tannakian category specified by generators, consider this definition of the <u>Mumford-Tate group of an abelian variety</u> (cf. [DMOS], pp. 39-47 and p. 63 f; see also 6.0 below):
Let $K = \overline{K}$ be an algebraically closed field, and $\sigma: K \to \mathbb{C}$. Let A be an abelian variety defined over K, and denote by $\langle A \rangle$ the smallest full tannakian subcategory of $\mathcal{M}_K$ containing $h^1(A)$ and $\mathbb{Q}(1)$. As $\otimes$-functor on $\langle A \rangle$ we take the restriction to $\langle A \rangle$ of $H_\sigma: \mathcal{M}_K \to \mathcal{V}ec_{\mathbb{Q}}$. Then $(\langle A \rangle, H_\sigma)$ corresponds to an affine group scheme $MT(A)$ over $\mathbb{Q}$, called the Mumford-Tate group of A.

From 2.3.2(d) we see that

$$MT(A) \hookrightarrow GL(H^1_\sigma(A)) \times \mathbb{G}_m .$$

But we know more: Deligne's fundamental theorem 2.1.3 implies that

$$H_\sigma : \langle A \rangle \to \mathcal{H}od_{\mathbb{Q}}$$

is a fully faithful functor to the category of rational Hodge structures - see 2.4.3 below. Thus, writing

$$V = H^1_\sigma(A) \quad \text{(as a } \mathbb{Q}\text{-Hodge structure)},$$

$$T^{a,b,m} = V^{\otimes a} \otimes \check{V}^{\otimes b} \otimes \mathbb{Q}(m) ,$$

for $a,b,m \in \mathbb{Z}$; $a,b \geq 0$, and calling Hodge cycles in $T^{a,b,m}$ those elements of $T^{a,b,m}$ that are pure of type $(0,0)$ in the Hodge decomposition of $T^{a,b,m} \otimes_{\mathbb{Q}} \mathbb{C}$, we find that $MT(A)$ is the $\mathbb{Q}$-algebraic subgroup of $GL(V) \times \mathbb{G}_m$ fixing all Hodge cycles in all spaces $T^{a,b,m}$.

The third description of $MT(A)$ is this: define $\mu : \mathbb{G}_m \to GL(V) \times \mathbb{G}_m$ over $\mathbb{C}$ by $\mu(z) = (\mu_1(z), z)$ and $\mu_1(z)(v) = (z \cdot v^{1,0}) + v^{0,1}$ $(z \in \mathbb{C}^*,\ v = v^{1,0} + v^{0,1} \in H^1_\sigma(A) \otimes \mathbb{C}.)$ Then $MT(A)$ is the smallest $\mathbb{Q}$-algebraic subgroup U of $GL(V) \times \mathbb{G}_m$ such that $U(\mathbb{C}) \supset \mu(\mathbb{C}^*)$.

$MT(A)$ is reductive. This follows alternatively from the existence of a polarization on A - which is fixed by $MT(A)$ because it defines a Hodge cycle in $T^{2,0,1}$, and the existence of which forces $MT(A)$ to have a compact real form - , or from the semisimplicity of $\langle A \rangle$, by 2.3.2(a). In fact the whole category of motives M_K is semisimple (i.e., every exact sequence in M_K splits), by [DMOS], II. 6.5, and $\langle A \rangle$ is a full tannakian subcategory of M_K.

Finally, $MT(A)$ is a torus if and only if all simple factors (up to isogeny) A_i of A admit complex multiplication by a CM-field E_i with $[E_i:\mathbb{Q}] = 2 \dim A_i$.

2.3.4 We have already quoted that, for any field K admitting an embedding $\sigma : K \to \mathbb{C}$, the category M_K of motives (for absolute Hodge cycles) over K is a semisimple tannakian category, equipped with the $\otimes$-functor $H_\sigma : M_K \to Vec_{\mathbb{Q}}$. The corresponding affine group scheme over $\mathbb{Q}$ is denoted $G(\sigma)$. It is proreductive, according to 2.3.2(a). $G(\sigma)$ as a whole looks prohibitively big and uncontrollable. In order to make it appear less outlandish, it is called the motivic Galois group. This terminology takes its clue from the classical son of $G(\sigma)$ to be discussed in the next paragraph.

2.4 SPECIAL MOTIVES (Reference: [DMOS], II § 6)

We shall need later on a few subcategories of M_K.

2.4.1 Artin motives: M_K^0

Let CV_K^0 be the subcategory of CV_K (2.2, step 1) formed by the h(X) with X a variety over K of dimension zero. For such an X, the $\overline{K}$-rational points $X(\overline{K})$ are just a finite set with a $Gal(\overline{K}/K)$-action, so consider the finite dimensional rational representation $\mathbb{Q}^{X(\overline{K})}$ of $Gal(\overline{K}/K)$, where we may view $Gal(\overline{K}/K)$ as a constant group scheme over K. In CV_K^0, one has

$$\begin{aligned} \mathrm{Hom}\,(h(X),h(Y)) &= C_{AH}^0(X\times Y) \\ &= \left(\mathbb{Q}^{X(\overline{K})\times Y(\overline{K})}\right)^{Gal(\overline{K}/K)} \\ &= \mathrm{Hom}_{Gal(\overline{K}/K)}(\mathbb{Q}^{X(\overline{K})}, \mathbb{Q}^{Y(\overline{K})}). \end{aligned}$$

Whence a fully faithful functor $CV_K^0 \to Rep_{\mathbb{Q}}(Gal(\overline{K}/K))$ into the tannakian category of finite-dimensional rational representations of $Gal(\overline{K}/K)$. Let M_K^0 be the smallest tannakian subcategory of M_K containing CV_K^0. Thus there is an equivalence of $\otimes$-categories between this category M_K^0 of (Emil) Artin motives and $Rep_{\mathbb{Q}}(Gal(\overline{K}/K))$.

For future reference, let us list the realizations of an Artin motive $M \in M_K^0$. We think of M as a representation of $Gal(\overline{K}/K)$, and denote by M_s the underlying finite-dimensional $\mathbb{Q}$-vector space.

For all $\sigma: K \hookrightarrow \mathbb{C}$, $H_\sigma(M) = M_s$

$$H_\sigma(M) \otimes \mathbb{C} = H^{0,0}$$

Hence for every prime number ℓ,

$$H_\ell(M) = M \otimes_{\mathbb{Q}} \mathbb{Q}_\ell \quad \text{(as } Gal(\overline{K}/K)\text{-module)}.$$

To determine the de Rham realization write M as Spec A, with $A = \Pi K_i$, where the $K_i \supset K$ are finitely many number fields. Now,

$$H_{DR}(\mathrm{Spec}\ K_i) = K_i \qquad \text{(as K-vector space)}$$
$$= (K_i \otimes_K \overline{K})^{\mathrm{Gal}(\overline{K}/K)}$$
$$= (\overline{K}^{\mathrm{Hom}_K(K_i,\overline{K})})^{\mathrm{Gal}(\overline{K}/K)}$$
$$= (\mathbb{Q}^{(\mathrm{Spec}\ K_i)(\overline{K})} \otimes_{\mathbb{Q}} \overline{K})^{\mathrm{Gal}(\overline{K}/K)}.$$

Therefore,

$$H_{DR}(M) = (M \otimes_{\mathbb{Q}} \overline{K})^{\mathrm{Gal}(\overline{K}/K)} .$$

2.4.2 Abelian varieties: $\mathcal{M}_K^{av}$

Let $\mathcal{M}_K^{av}$ be the tannakian subcategory of $\mathcal{M}_K$ generated by motives of abelian varieties and Artin motives over K. Since all of $h(X)$ is given by the exterior algebra of $h^1(X)$, for an abelian variety X, $\mathcal{M}_K^{av}$ is already generated by the $h^1(X)$ and $\mathcal{M}_K^0$.

2.4.3 Theorem. *If K is algebraically closed, and $\sigma: K \to \mathbb{C}$ is any embedding, then the $\otimes$-functor*

$$H_\sigma : \mathcal{M}_K^{av} \to \mathcal{H}od_{\mathbb{Q}}$$

into the tannakian category of rational Hodge structures is fully faithful.

This is an easy reformulation of Delgine's theorem 2.1.3 above. Let us give a proof that would work for any category C of motives generated by varieties of which one could prove that, over algebraically closed fields, every Hodge cycle on them was absolutely Hodge: We have to show that any $\mathbb{Q}$-linear map $f_\sigma: H_\sigma(M) \to H_\sigma(N)$, for $M,N \in C$, which over $\mathbb{C}$ respects the Hodge decompositions comes from an "absolute Hodge cycle on $M \times N$". By the comparison isomorphisms f_σ induces a $\mathbb{Q}_{\mathbb{A}^f}$-linear map $f_{\mathbb{A}^f}: H_{\mathbb{A}^f}(M/K) \to H_{\mathbb{A}^f}(N/K)$, and a $\mathbb{C}$-linear map $f_{DR,\mathbb{C}}: H_{DR}(M/\mathbb{C}) \to H_{DR}(N/\mathbb{C})$ respecting the Hodge filtrations. There is a field $L \supset K$, say $L = \overline{L}$, of finite transcendence degree over

K, and an extension $\tilde{\sigma}$ of σ from K to L such that $f_{DR,\mathbb{C}}$ is already defined over $L^{\tilde{\sigma}}$. Then $(f_{DR,L}, f_{\mathbb{A}}f)$ is a Hodge cycle on $M \times N$ over L relative to $\tilde{\sigma}$. By assumption on C, as L is algebraically closed it is an absolute Hodge cycle. Now Proposition 2.1.2(a) shows that it can already be defined over $\overline{K} = K$.

Certain classes of algebraic varieties are known to have motives isomorphic in $\mathcal{M}_K$ to objects of $\mathcal{M}_K^{av}$. E.g., curves (via their jacobians), but also K3-surfaces and Fermat varieties. We shall recall these results as we need them.

Since our main concern is with algebraic Hecke characters we are eventually going to concentrate on the subcategory of $\mathcal{M}_K^{av}$ generated by abelian varieties with (potential) complex multiplication (and Artin motives). First, however, we have to explain how motives can be related to algebraic Hecke characters.

3. Motives of rank 1

3.0 The notion of complex multiplication for abelian varieties generalizes to motives in the following way. Let K be a field embeddable into $\mathbb{C}$ and E a number field of finite degree over $\mathbb{Q}$. The category $\mathcal{M}_K(E)$ of motives over K with coefficients in E has objects the pairs (M,θ), with M a motive over K (i.e., an object of $\mathcal{M}_K$), and $\theta : E \to \mathrm{End}(M)$ an embedding of $\mathbb{Q}$-algebras. The morphisms in $\mathcal{M}_K(E)$ are the obvious ones, respecting the E-structures. $\mathcal{M}_K(E)$ is a $\otimes$-category via

$$(M,\theta) \otimes_E (M',\theta') = (N,\iota)$$

where N is the direct factor of $M \otimes M'$ on which

$$\begin{array}{ccc} E \overset{\theta}{\hookrightarrow} \mathrm{End}(M) & \hookrightarrow & \\ & & \mathrm{End}(M \otimes M') \\ E \overset{\theta'}{\hookrightarrow} \mathrm{End}(M') & \hookrightarrow & \end{array}$$

agree to define ι. For an alternative description of $\mathcal{M}_K(E)$, see [DP],2.1, "<u>langage</u> B".

The E-structure $\theta: E \to \mathrm{End}(M)$ defines E-module structures on all the realizations of a motive (M,θ) in $M_K(E)$. Thus, for $\sigma: K \to \mathbb{C}$, $H_\sigma(M)$ is an E-rational Hodge structure, i.e., an E-vector space with a decomposition of $E \otimes \mathbb{C}$-modules

$$H_\sigma(M) \otimes_{\mathbb{Q}} \mathbb{C} = \bigoplus_{p,q} H^{pq} \quad ,$$

such that complex conjugation interchanges H^{pq} and H^{qp}. (Cf. 6.0/1) Forgetting this Hodge decomposition of $H_\sigma(M)$, the pair $(M_K(E), H_\sigma)$ is a neutralized tannakian category <u>over</u> E, and the corresponding E-group scheme is $G(\sigma) \times_{\mathbb{Q}} E$, the motivic Galois group considered over E. (This is most easily seen in "<u>langage</u> B" quoted above.) In other words, there is an equivalence of categories induced by H_σ,

$$M_K(E) \to Rep_E(G(\sigma)/E).$$

<u>The rank of</u> (M,θ) <u>in</u> $M_K(E)$ is defined to be the trace of identity in the corresponding representation:

$$rk(M,\theta) = rk_E M = \dim_E H_\sigma(M).$$

(This is, of course, independent of σ.)

The de Rham realization of a motive (M,θ) - or, as we shall simply write, M - in $M_K(E)$ is a filtered $E \otimes K$-module, free of rank $rk_E M$.

For all prime numbers $\ell, H_\ell(M)$ is a free $E \otimes \mathbb{Q}_\ell$-module of rank $rk_E M$, with an $E \otimes \mathbb{Q}_\ell$-linear action of $\mathrm{Gal}(\bar{K}/K)$. Since $E \otimes \mathbb{Q}_\ell = \prod_{\lambda | \ell} E_\lambda$, λ ranging over the places of E dividing ℓ, there is a decomposition of $\mathrm{Gal}(\bar{K}/K)$-representations

$$H_\ell(M) = \bigoplus_{\lambda|\ell} H_\lambda(M),$$

the $H_\lambda(M) = H_\ell(M) \otimes_{E\otimes\mathbb{Q}_\ell} E_\lambda$ being <u>the</u> λ-<u>adic Galois representations of</u> M.

The realizations of $M \otimes_E M'$ are simply the E-linear tensor products of the realizations of $M, M' \in \mathcal{M}_K(E)$.

The functors on $\mathcal{M}_K$: extension and restriction of the base field K clearly induce functors on $\mathcal{M}_K(E)$ - cf. 2.2.2. In addition, if $E \subset E'$ is a (necessarily finite) extension, there are functors

$$\otimes_E E' : \mathcal{M}_K(E) \to \mathcal{M}_K(E')$$

$$|_E : \mathcal{M}_K(E') \to \mathcal{M}_K(E)$$

of <u>extension and restriction of the field of coefficients.</u> $|_E$ is simply forgetting the E'-action, except for that of E. - $\otimes_E E'$ sends M into $M \otimes_E E'$, where $E' \in \mathcal{M}_K$ is the first component of the unit object (E', θ') of $\mathcal{M}_K(E')$.

<u>Assume now that</u> K <u>(as well as</u> E) <u>is a number field of finite degree over</u> $\mathbb{Q}$.

<u>3.1 Proposition</u> (cf. [DP], 8.1 (iii)) <u>Let</u> $M \in \mathcal{M}_K(E)$ <u>with</u> $rk_E M = 1$. <u>If the system of the</u> $H_\lambda(M)$, <u>for all finite places</u> λ <u>of</u> E, <u>is a strictly compatible system of</u> E-<u>rational</u> λ-<u>adic</u> <u>Galois-representations</u> χ_λ over K, <u>then there is an algebraic Hecke character</u> χ <u>of</u> K <u>with values in</u> E <u>such that for all</u> λ <u>and almost all primes</u> $\mathfrak{p}$ <u>of</u> K, $\chi(\mathfrak{p}) = \chi_\lambda(\mathrm{Frob}\ \mathfrak{p})$.

This is just an application of Proposition 1.4 above.

<u>3.2 Remark:</u> Absolute Hodge cycles do not lend themselves to reduction mod $\mathfrak{p}$ - or at least, we do not know how to prove that they do. This is why it is not even known, for a general motive M in $\mathcal{M}_K$, that the $H_\ell(M)$, for all rational primes ℓ, form a compatible system of rational ℓ-adic representations. This is true for $M \in \mathcal{CV}_K$ by the "Weil conjectures", but it cannot be shown to carry over to all motives constructed in Step 2 of the construction of $\mathcal{M}_K$. - On the positive side however, it will be shown in § 6 that every rank 1 motive in $\mathcal{M}_K^{av}(E)$ has strictly compatible λ-adic representations, and therefore defines a Hecke character. So, Proposition 3.1. will be shown to be far from vacuous.

We have now motivated the basic notion in the "geometric" theory of algebraic Hecke characters:

3.3. Definition. Let χ be an algebraic Hecke character of K with values in E. A motive $M \in M_K(E)$ is said to be a motive for χ, if $rk_E M = 1$ and for all finite places λ of E, and all prime ideals $\mathfrak{p}$ of K with $\mathfrak{p} \nmid f_\chi \cdot N\lambda$, the λ-adic representation $H_\lambda(M)$ of $Gal(\bar{K}/K)$ is unramified at $\mathfrak{p}$ and a geometric Frobenius element $Frob\,\mathfrak{p} \in Gal(\bar{K}/K)$ acts on $H_\lambda(M)$ via multiplication by $\chi(\mathfrak{p})$.

In other words, M is a motive for χ, if $H_\lambda(M) = \chi_\lambda$, in the notation of 0 § 5.

The typical example of a motive for an algebraic Hecke character is an abelian variety with complex multiplication - see § 1.

4. A standard motive for a Hecke character

Let K be embeddable into $\mathbb{C}$. Let CM_K be the Tannakian subcategory of M_K (equivalently: of M_K^{av}) generated by the Artin motives over K and by the motives $h^1(A)$, where A is an abelian variety over K which, over $\bar{K}$, has complex multiplication (in the sense that $End_{\bar{K}} A$ contains a number field E of degree $[E:\mathbb{Q}] = 2 \dim A$.) Given a number field E (of finite degree over $\mathbb{Q}$), we can consider the category $CM_K(E)$ of motives M in CM_K that are equipped with an E-action, $E \to End(M)$.

4.1 Theorem. Suppose K is a number field. For any algebraic Hecke character χ of K with values in E, there exists a motive $M(\chi) \in CM_K(E)$ which is a motive for χ, in the sense of 3.3.

Elementary and direct proof of 4.1

4.1.0 If χ is of the form $\mu . \mathbb{N}^{w/2}$, for a character of finite order μ on $Gal(\bar{K}/K)$ with values in E^*, then we can write down a motive for χ in $CM_K(E)$ like this:

$$M(\chi) = [\mu] \otimes_E E(\tfrac{-w}{2}) ,$$

where $[\mu] \in M_K^0(E)$ is the rank 1 Artin motive for μ with coefficients in E (i.e., $s \in \mathrm{Gal}(\bar{K}/K)$ acts on E via multiplication by $\mu(s) \in E^*$), and $E(\frac{-w}{2}) = \mathbb{Q}(\frac{-w}{2}) \otimes_{\mathbb{Q}} E$, E being here the unit object in $M_K(E)$.

Thus calling K' the field of all numbers of CM-type in K, let us henceforth <u>assume</u>, without loss of generality, <u>that</u> K' <u>and</u> E <u>are</u> <u>CM-fields</u> - cf. **0** § 3.

<u>4.1.1</u> We now treat the case that the infinity-type of χ is of the form $T = \Phi' \circ N_{K/K'}$ with a CM-type Φ' of K', i.e., χ is of weight -1 and the invariants $n(\sigma,\tau)$, for $\sigma \in \mathrm{Hom}(K,\mathbb{C})$, $\tau \in \mathrm{Hom}(E,\mathbb{C})$, introduced in **0** § 4 are all either -1 or 0. Exchanging the rôles of K and E these same $n(\sigma,\tau)$ also define a CM-type Φ of the field

$$E_\circ = \mathbb{Q}(K'^{\Phi'}) = \mathbb{Q}(K^T) \subset E ,$$

called the <u>reflex</u> <u>type</u> of Φ'. By a theorem of Casselman, [ShiL] , Theorem 6, there is an abelian variety A defined over K with complex multiplication by the ring of integers of E, of CM-type $(E, \Phi \circ N_{E/E_\circ})$, such that $h_1(A)$ <u>is</u> <u>a</u> <u>motive</u> <u>for</u> χ. - In fact, if one tries to be very neat, A may be constructed as being a direct summand in $h_1(R_{L|K}B)$, for a suitable (abelian) extension L of K such that $\chi \circ N_{L|K}$ takes values in $E_\circ^*$, and B , provided by Casselman, an abelian variety over L of CM-type $(E_\circ, \Phi)$ with character $\chi \circ N_{L|K}$ - cf. [GS], théorème 4.1.

<u>4.1.2</u> All infinity-types of algebraic Hecke characters of K are $\mathbb{Z}$-linear combinations of the CM-types discussed in 4.1.1. - This is an easy exercise, observing the homogeneity condition $n(\sigma,\tau) + n(c\sigma,\tau) = w$, cf. **0** §§ 3,4.

<u>4.1.3</u> We can now start on the general case of 4.1 (always under the assumption that K' is a CM-field). Write the infinity-type T of χ as

$$T = \prod_i T_i^{n_i} \quad ,$$

with $n_i \in \mathbf{Z}$ and T_i CM-types like in 4.1.1. There is a finite extension field $E' \supset E$ such that, for all i, there exists an algebraic Hecke character χ_i of K with values in E' of infinity-type T_i - see **0** § 3. Let A_i be an abelian variety attached to χ_i as in 4.1.1. Put

$$M' = (\otimes_{E'_i} h_1(A_i)^{\otimes_{E'} n_i}) \otimes_{E'} [\mu]$$

where $\mu = \chi(\Pi\chi_i^{-n_i})$ is of finite order, and $[\mu]$ is the Artin motive for μ in $M_K^0(E')$. Then M' is a motive for χ, if we consider χ <u>to take values in</u> E', rather than E. Thus it remains to "descend the coefficients".

<u>4.1.4</u> There is a finite (abelian) extension L of K such that all characters χ_i used in 4.1.3 take their values in E^* when composed with $N_{L|K}$. Therefore, taking, for simplicity, L such that every $\chi_i \circ N_{L|K}$ takes its values in the corresponding reflex-field $E_{\circ,i}$ (see 4.1.1), we see that $M' \times_K L$ is of the form $M_L \otimes_E E'$, with M_L a direct factor of the motive:

$$\{\otimes_{E_i} [(h_1(B_i)^{\otimes_{E_{\circ,i}} n_i}) \otimes_{E_{\circ,i}} E]\} \otimes_E [\mu \circ N_{L|K}]_E \quad ,$$

the B_i being as B in the last sentence of 4.1.1. So, M_L is a motive for $\chi \circ N_{L|K}$ in $CM_L(E)$. In other words, there is a projector (an absolute Hodge cycle) $\pi \in \mathrm{End}_{M_L}(M' \times_K L)$ carving out the E-structure M_L. We have to show that π is already defined over K. Since it is an absolute Hodge cycle it is enough to show that its $\mathbf{A}^f$-component is invariant under $\mathrm{Gal}(\bar{K}/K)$. But M' is a motive for the character χ which takes values in E. So π cannot possibly be affected by the action of $\mathrm{Gal}(\bar{K}/K)$ on $H_{\mathbf{A}^f}(M')$.

<u>4.2 Remark.</u> In view of 1.7 above, the motive $M(\chi)$ for χ which was just constructed has its Hodge structure determined by the infinity-type of χ. Explicitly, for all $\sigma: K \to \mathbb{C}, \tau: E \to \mathbb{C}$, one finds

$$H_\sigma(M(\chi)) \otimes_{E,\tau} \mathbb{C} \subset H^{n(\sigma,\tau),w-n(\sigma,\tau)} \quad .$$

(See **0** § 4 for the notation.)

The proof of theorem 4.1 which we have presented is "elementary and direct" in that it starts immediately from the geometry of the varieties that generate CM_K, and does <u>not</u> use Deligne's theorem 2.1.3 about Hodge cycles on abelian varieties. It does use Casselman's theorem, i.e., the Shimura-Taniyama reciprocity law for abelian varieties with complex multiplication.

Using 2.1.3 it is possible to gain much more insight into the structure of CM_K, reproving theorem 4.1 (via the interpretation of algebraic characters given in **0** § 7) and generalizing the Shimura-Taniyama reciprocity law. Specifically, what one has to do is to identify $CM_{\mathbb{Q}}$ with $\mathcal{R}ep_{\mathbb{Q}}(\mathfrak{T})$, for the Taniyama group $\mathfrak{T}$. We shall sketch this in § 6 below, thereby obtaining additional information about all rank 1 motives constructed from abelian varieties.

5. Unicity of M(χ)

Let K and E be number fields of finite degree over $\mathbb{Q}$, and consider the category of motives $M_K^{av}(E)$. These are the motives in M_K^{av} - see 2.4.2 - with an E-action. Since CM_K is a subcategory of M_K^{av}, there always exists - by theorem 4.1 - a motive $M(\chi)$ in $M_K^{av}(E)$ for a given algebraic Hecke character χ of K with values in E.

5.1 Theorem. *Up to isomorphism, there is only one motive* $M(\chi)$ *in* $M_K^{av}(E)$ *for a given algebraic Hecke character* χ *of* K *with values in* E.

Proof. Let M in $M_K^{av}(E)$ be any motive for χ. For any $\sigma: K \hookrightarrow \mathbb{C}$, $H_\sigma(M)$ is a Hodge structure of rank 1 over E. In particular, it is indecomposable and therefore pure of some weight w (cf. 6.0 below). The relations

$$H_\sigma(M) \otimes_{E,\tau} \mathbb{C} \subset H^{n(\sigma,\tau),w-n(\sigma,\tau)}$$

define invariants $n(\sigma,\tau)$ for all $\tau:E\to\mathbb{C}$. They actually satisfy $n(\alpha\sigma,\alpha\tau) = n(\sigma,\tau)$, for all $\alpha\in \mathrm{Aut}\,\mathbb{C}$ (because E operates on M through absolute Hodge cycles.) So by 0 §§ 3 and 4, there is some algebraic Hecke character Ψ of some number field $L\supset K$ with values in E having the $n(\sigma,\tau)$'s as its invariants. Let $M(\Psi)\in CM_L(E)$ be the motive for Ψ constructed in 4.1. By remark 4.2 the Hodge structure $H_{\tilde\sigma}(M(\Psi))$, for every embedding $\tilde\sigma:L\to\mathbb{C}$ extending $\sigma:K\hookrightarrow\mathbb{C}$, is E-compatibly isomorphic to $H_\sigma(M)$. By theorem 2.4.3, M and $M(\Psi)$ are isomorphic in $M_{\bar K}^{av}(E)$. In view of 2.1.2(b), they are isomorphic over some finite extension L' of L. Recalling that M was a motive for χ, and $M(\Psi)$ for Ψ, we find that $\chi\circ N_{L'/K} = \Psi\circ N_{L'/L}$. Hence the $n(\sigma,\tau)$ which by construction describe the infinity-type of Ψ are also the invariants attached to the character χ. Thus we have shown that, if M is an arbitrary motive for χ in $M_K^{av}(E)$, its Hodge realizations $H_\sigma(M)$ are those determined by (the infinity-type of) χ, as in 4.2. This establishes an isomorphism (an absolute Hodge cycle) between M and our standard motive $M(\chi)$ over $\bar K$. But $H_{\mathbb{A}^f}(M)$ and $H_{\mathbb{A}^f}(M(\chi))$ are isomorphic $\mathrm{Gal}(\bar K/K)$-representations by definition. So the isomorphism is defined over K.

q.e.d.

5.2 So far we have verified all but part (iii) of conjecture 8.1 in [DP] for the category M_K^{av}. (In fact, our description in 4.2 of the Hodge decomposition of $M(\chi)$ is equivalent to the characterization of the Hodge filtration given by Deligne in [DP], 8.1 (iv).) As to [DP], 8.1 (iii), it will be shown in § 6 that every motive M in $CM_K(E)$ has strictly compatible E-rational λ-adic representations $H_\lambda(M)$. By proposition 3.1, this will settle [DP], 8.1 (iii) for the category CM_K. But in fact, it will automatically take care also of the rank 1 motives in M_K^{av}:

5.3 Remark Every motive of rank 1 in $M_K^{av}(E)$ is isomorphic to a motive of $CM_K(E)$.

Proof. Let M be in $M_K^{av}(E)$; let $\sigma:K\hookrightarrow\mathbb{C}$, and assume that $\dim_E H_\sigma(M) = 1$. Then $H_\sigma(M)$ is an E-rational Hodge structure of the kind described in 4.2 and the proof of 5.1. It occurs as $H_{\tilde{\sigma}}(N)$, for some N in $CM_L(E)$, for some number field L with $K \subset L \subset \overline{K}$ and $\tilde{\sigma}$ extending σ to L. By 2.4.3, $M \times_K \overline{K}$ is isomorphic to a motive in $CM_{\overline{K}}(E)$. This being true over some finite extension L' of K (and L), we see that M is isomorphic to a motive in $CM_K(E)$. (E.g., M occurs as a direct factor in $R_{L'/K}N$.)

6. Representations of the Taniyama group

The affine group scheme over $\mathbb{Q}$ corresponding, by tannakian philosophy, to the neutralized category of motives $(CM_{\mathbb{Q}}, H_B)$ - see § 4 above for the notation - is (isomorphic to) the Taniyama group introduced by Langlands in [Lg], 5. This fact was first proved by Deligne: see [DMOS], IV. Using a formalism of Tate's completed by an argument of Deligne - see [LCM], chap. 7 - , the proof can be given much more explicitly. This second proof is certainly part of the folklore on this subject - I myself am indebted to G. Anderson for explaining it to me - and J.S. Milne is preparing a book which will contain it in detail. In this section we shall give an extremely sketchy account of how this proof proceeds, and then apply the theorem to settle the only question left open in 5.2. The results of this section will not be substantially used in the sequel. They are however, essential for G. Anderson's formalism (section 7), and they complete the picture we are drawing of motives for Hecke characters. - The first two subsections: 6.0 and 6.1, are more detailed than the rest because they give a more thorough basis to things that have been used before: rational Hodge structures and the Serre group. The definition of CM Hodge structures in 6.1 was suggested by R. Pink.

6.0 Rational Hodge structures

6.0.0 A rational Hodge structure of weight w is a finite dimensional $\mathbb{Q}$-vector space V equipped with a decomposition

$$V \otimes \mathbb{C} = \bigoplus_{\substack{p+q=w \\ p,q\in\mathbf{Z}}} V^{pq}$$

such that $(1 \otimes c)\ V^{pq} = V^{qp}$, for c = complex conjugation. A rational Hodge structure is a finite direct sum of rational Hodge structures of fixed weights. A homomorphism of rational Hodge structures V_1, V_2 is a $\mathbb{Q}$-linear map $f: V_1 \to V_2$ such that, for all $p,q \in \mathbf{Z}$, one has

$$(f \otimes 1_{\mathbb{C}})\ V_1^{pq} \subset V_2^{pq}\ .$$

6.0.1 Reformulated in a tannakian way, the extra structure on the $\mathbb{Q}$-vector space V amounts to a representation $h: \mathbb{S} \to GL(V)$ defined over $\mathbb{R}$, where $\mathbb{S} = R_{\mathbb{C}/\mathbb{R}}\mathbb{G}_m$ - see [DeH II], 2.1. The translation is given by the rule

$$h(z)\big|_{V^{pq}} = \text{multiplication by } z^p\bar{z}^q\ ,$$

for $z \in \mathbb{C}^* = \mathbb{S}(\mathbb{R})$.

The inclusion $\mathbb{R}^* \hookrightarrow \mathbb{C}^*$ gives rise to a canonical map $\mathring{w}: \mathbb{G}_m \to \mathbb{S}$ over $\mathbb{R}$. Given a rational Hodge structure V, we set $w = h \circ \mathring{w} : \mathbb{G}_m \to GL(V)$. If V is of weight n, then $w(\lambda)$ acts as multiplication by λ^n on V; this justifies the letter w, and implies that, among the "real Hodge structures" $h: \mathbb{S} \to GL(V)/\mathbb{R}$, the rational Hodge structures are precisely those for which w is defined over $\mathbb{Q}$. Over $\mathbb{C}$, denote by $\mu: \mathbb{G}_m \to GL(V)/\mathbb{C}$ the complex cocharacter given by the rule

$$\mu(z)\big|_{V^{pq}} = \text{multiplication by } z^p\ ,$$

for $z \in \mathbb{C}^* = \mathbb{G}_m(\mathbb{C})$. Its (imagewise) complex conjugate

$$\bar{\mu}: z \mapsto (v^{pq} \mapsto \bar{z}^q \cdot v^{pq})$$

is algebraic (not over $\mathbb{C}$ but) over $\mathbb{R}$, and $(\mu\bar{\mu})$ takes values in $GL(V \otimes \mathbb{R})$ on $\mathbb{C}^*$. Thus it defines an algebraic homomorphism $\mathbb{S} \to GL(V)$ over $\mathbb{R}$, which is none other than h. Either h or μ suffice to characterize the Hodge structure on a given $\mathbb{Q}$-vector space V. This is convenient, for instance, in defining the tensor product of rational Hodge structures via the tensor product of real $\mathbb{S}$-representations. - Rational Hodge structures form a $\mathbb{Q}$-linear $\otimes$-category.

6.0.2 Define the Mumford-Tate group of a rational Hodge structure V by generalizing 2.3.3: MT(V) is the $\mathbb{Q}$-algebraic subgroup of $GL(V) \times \mathbb{G}_m$ that fixes all elements of pure type (0,0) in all spaces of the form

$$T^{a,b,m} = V^{\otimes a} \otimes \check{V}^{\otimes b} \otimes \mathbb{Q}(m),$$

where $a,b,m \in \mathbb{Z}$; $a,b \geq 0$; T being viewed as tensor product of the natural representations of GL(V) on V and $\check{V}$, and the representation: multiplication by λ^{-1} of $\mathbb{G}_m$ on $\mathbb{Q}(1)$. Equivalently ([DMOS],I. 3.4.), MT(V) is the smallest $\mathbb{Q}$-algebraic subgroup of $GL(V) \times \mathbb{G}_m$ such that $MT(V)(\mathbb{C})$ contains the image of

$$\begin{aligned} \mathbb{C}^* &\to GL(V \otimes \mathbb{C}) \times \mathbb{C}^* \\ z &\to (\mu(z), z) \end{aligned} .$$

It is often convenient to identify MT(V) with its first projection. Thus MT(V) becomes the smallest $\mathbb{Q}$-algebraic subgroup of GL(V) the $\mathbb{C}$-rational points of which contain the image of μ. The first description then runs:

$$MT(V)(\mathbb{Q}) = \left\{ \gamma \in GL(V) \;\middle|\; \begin{array}{l} \text{for all } a,b,m \text{, and all} \\ t \in V^{\otimes a} \otimes \check{V}^{\otimes b} \cap (V^{\otimes a} \otimes \check{V}^{\otimes b})^{m,m}, \\ \text{there is } \lambda \in \mathbb{Q}^* \text{ such that } \gamma(t) = \lambda^m t \end{array} \right\}$$

6.0.3 As $MT(V)(\mathbb{C})$ receives the cocharacter μ, one also has

$$h: \mathbb{S} \to MT(V) \quad \text{over} \quad \mathbb{R},$$

and

$$w: \mathbb{G}_m \to MT(V) \quad \text{over} \quad \mathbb{Q}.$$

μ and w have more or less surfaced already in chapter 0, 7.3.4 , and we are now going to reconsider the Serre group Z in the context of rational Hodge structures of CM type.

6.1 CM Hodge Structures (cf. also [DMOS], III. 1)

6.1.0 Definition: Let V be a rational Hodge structure, $MT(V)$ its Mumford-Tate group (6.0.2) and $w:\mathbb{G}_m \to MT(V)/\mathbb{Q}$ the associated cocharacter (6.0. 1/3). V is called a CM Hodge structure if $MT(V)$ is a torus and $(MT(V)/w(\mathbb{G}_m))(\mathbb{R})$ is compact.

CM Hodge structures form a $\otimes$-subcategory of all rational Hodge structures. We are going to show that the Mumford-Tate group of a CM Hodge structure is a quotient of the Serre group Z, introduced in **0** , 7.3.3. Recall that $Z = \varprojlim Z_K$, where K runs over number fields and Z_K is the quotient of $R_{K/\mathbb{Q}}\mathbb{G}_m$ by a sufficiently small arithmetic subgroup. Observe that the maps $\tilde{w}$ und $\tilde{\mu}$ defined in **0**, 7.3.4 actually give rise to cocharacters of $R_{K/\mathbb{Q}}\mathbb{G}_m$ (defined over $\mathbb{R}$, and $\mathbb{C}$, resp.), for all number fields K (embedded into $\overline{\mathbb{Q}}$, as explained in **0**, 7.3.2).

6.1.1 Lemma *Let V be a rational Hodge structure such that $MT(V)$ is a torus.*

(i) *For any sufficiently large number field* $K \hookrightarrow \overline{\mathbb{Q}}$, *there exists a unique homomorphism*

$$\nu: R_{K/\mathbb{Q}}\mathbb{G}_m \to MT(V)$$

of $\mathbb{Q}$-algebraic groups rendering the following diagram commutative.

$$\begin{array}{ccc} R_{K/\mathbb{Q}}\mathbb{G}_m \times \mathbb{C} & \xrightarrow{\nu \times \mathbb{C}} & MT(V) \times \mathbb{C} \\ & {}_{\tilde{\mu}}\nwarrow \quad \nearrow_{\mu} & \\ & \mathbb{G}_m/\mathbb{C} & \end{array}$$

(ii) *For* $K \subset L$, *the maps* ν *are compatible via* $N_{L/K}$.

(iii) ν *is faithfully flat.*

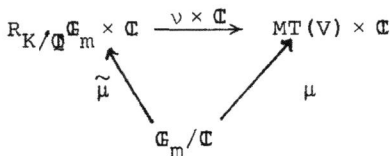

Proof. (i) Translate into a statement on character groups; the requirement that ν be defined over $\mathbb{Q}$ then forces (for K normal over $\mathbb{Q}$ splitting $MT(V)$):

$$\nu^*(f) = \sum_{\sigma\in G(K/\mathbb{Q})} \mu^*(f^{(\sigma^{-1})}) \cdot \sigma \in X(R_{K/\mathbb{Q}}\mathbb{G}_m),$$

for $f \in X(MT(V))$.

(ii) follows from the uniqueness of ν.
(iii) expresses the fact that μ "generates" the $\mathbb{Q}$-algebraic group $MT(V)$.

6.1.2 Remark Care has to be taken with the galois action(s) on Z: cf. [Lg], p. 219/20; [DMOS], III (1.3), (1.8); [DMOS], IV, (B)... Since we just used a galois invariance in the proof of (i), let us make explicit our setup: we define a left action of $G(\overline{\mathbb{Q}}/\mathbb{Q})$ on

$$R_{K/\mathbb{Q}}\mathbb{G}_m(\overline{\mathbb{Q}}) = (\overline{\mathbb{Q}}^*)^{\mathrm{Hom}(K,\overline{\mathbb{Q}})} \quad \text{by the rule}$$

$$((z_\sigma)_{\sigma\in \mathrm{Hom}(K,\overline{\mathbb{Q}})})^s = (z_{s^{-1}\circ\sigma})_\sigma, \quad \text{for } s \in G(\overline{\mathbb{Q}}/\mathbb{Q})$$

This transports to a left action on characters

$$f: R_{K/\mathbb{Q}}\mathbb{G}_m \to \mathbb{G}_m/\overline{\mathbb{Q}} \quad \text{via}: f^s((z_\sigma)) = f((z_\sigma)^{s^{-1}}).$$

Identifying f with $\sum_\sigma n_\sigma \sigma \in \mathbb{Z}[\mathrm{Hom}(K,\overline{\mathbb{Q}})]$, one finds

$$(\sum_\sigma n_\sigma \sigma)^s = \sum_\sigma n_{s^{-1}\circ\sigma}\sigma.$$

This is the action we have used on characters of $R_{K/\mathbb{Q}}\mathbb{G}_m$ (cf. $\mathbf{0}$, §§ 2 und 4). Passing to the quotient Z_K and to the limit Z, this yields, for a character $f: G(\overline{\mathbb{Q}}/\mathbb{Q}) \to \mathbb{Z}$ of Z (as in $\mathbf{0}$. 7.3.4): $f^s(t) = f(s^{-1}t)$, for $s,t \in \mathrm{Gal}(\overline{\mathbb{Q}}/\mathbb{Q})$. (Note that $f^{(st)} = (f^t)^s$.)

There is, of course, another left action of $\mathrm{Gal}(\overline{\mathbb{Q}}/\mathbb{Q})$ on $X(Z)$, namely by right translation:

$$(s \cdot f)(t) = f(ts).$$

In the preceding setup, it is induced by the rule $(z_\sigma)^s_\sigma = (z_{\sigma s})_\sigma$ on $R_{K/\mathbb{Q}}\mathbb{G}_m(\overline{\mathbb{Q}})$ - where K has now to be assumed normal over $\mathbb{Q}$ (and always embedded into $\overline{\mathbb{Q}}$). - This second action will become relevant as of 6.2.0.

<u>6.1.3 Proposition [R. Pink]</u> <u>Let</u> V <u>be a rational Hodge structure such that</u> $MT(V)$ <u>is a torus. Choose</u> K <u>so that</u> ν <u>exists as in 6.1.1. Then the following are equivalent.</u>

(i) V <u>is a CM Hodge structure.</u>

(ii) $\ker \nu$ <u>contains an arithmetic subgroup.</u>

(iii) <u>The subgroup</u> $X(MT(V)) \overset{\nu^*}{\hookrightarrow} X(R_{K/\mathbb{Q}}\mathbb{G}_m)$ <u>is contained in</u> $X(Z_K)$.

<u>Proof.</u> (i) ⇒ (ii). Assume (without loss of generality) that K is totally imaginary, say $[K:\mathbb{Q}] = 2r$, and consider the diagram

$$\begin{array}{ccccc} \mathbb{S}^r & = & R_{K/\mathbb{Q}}\mathbb{G}_m \times \mathbb{R} & \xrightarrow{\nu\times\mathbb{R}} & MT(V)\times\mathbb{R} \\ \mathring{w}\uparrow\ \downarrow N_{\mathbb{C}/\mathbb{R}} & & \tilde{w}\nwarrow & & \nearrow w\times\mathbb{R} \\ (\mathbb{G}_m/\mathbb{R})^r & & & \mathbb{G}_m/\mathbb{R} & \end{array}$$

whose commutativity is easily checked using 1.6.1 (i). According to Dirichlet, the units in $R_{K/\mathbb{Q}}\mathbb{G}_m$ map to $(\mathbb{G}_m/\mathbb{R})^r$ with a finite kernel. As $N_{\mathbb{C}/\mathbb{R}} \circ \mathring{w}(x) = x^2$, any sufficiently small arithmetic subgroup U of $R_{K/\mathbb{Q}}\mathbb{G}_m$ is contained in $\mathring{w}(\mathbb{G}_m/\mathbb{R})^r$, and since $(MT(V)/w(\mathbb{G}_m))(\mathbb{R})$ is compact we can achieve that $\nu(U)(\mathbb{R}) \subset w(\mathbb{G}_m/\mathbb{R})$. Now assume furthermore that $U \subset \ker N_{K/\mathbb{Q}}$, the norm-1-subgroup of $R_{K/\mathbb{Q}}\mathbb{G}_m$. Then $U(\mathbb{R}) \cap \tilde{w}(\mathbb{G}_m/\mathbb{R})$ is obviously finite. Thus, in view of the commutative triangle above, we conclude that $U \subset \operatorname{Ker} \nu$, provided again that U is sufficiently small.

(ii) and (iii) just express the two possible definitions of Z_K - as quotient of $R_{K/\mathbb{Q}}\mathbb{G}_m$, or via its character group.

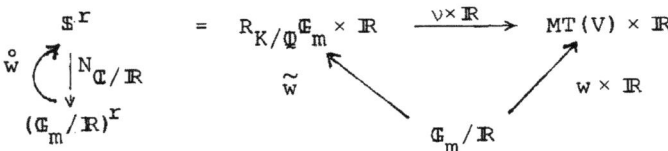

Finally, assuming (iii), we find (cf. **0**, 7.3.2)

$$X(MT(V)/w(\mathbb{G}_m)) \subset \{\lambda \in X(R_{K/\mathbb{Q}}\mathbb{G}_m) \mid \begin{array}{l}\lambda(\sigma) = -\lambda(c\sigma) \\ \text{for all } \sigma: K \hookrightarrow \overline{\mathbb{Q}}\end{array}\}$$

So, c acts as -1 on $X(MT(V)/w(\mathbb{G}_m))$, and $(MT(V)/w(\mathbb{G}_m))(\mathbb{R})$ is a quotient of

$$\ker (N_{\mathbb{C}/\mathbb{R}}: \mathbb{S} \to \mathbb{G}_m)(\mathbb{R})^r ,$$

that is, a quotient of a product of S^1's, and therefore compact.

q.e.d.

<u>6.1.4 Corollary</u> <u>The ⊗-category of all CM Hodge structures, together with the functor which, to a CM Hodge structure, associates its underlying $\mathbb{Q}$-vector space is a neutralized tannakian category over $\mathbb{Q}$, with the Serre group</u> Z <u>(see</u> **0**, 7.3.3) <u>as corresponding affine group scheme.</u>

It is clear how the mapping ν of 6.1.1 defines a representation of Z if the conditions of 6.1.3 are satisfied. So the proof of 6.1.4 is obvious, but let us illustrate the corollary in our principal

<u>6.1.5 Example.</u> Let χ be an algebraic Hecke character of K with values in E, and let $M(\chi) \in CM_K(E)$ be the standard motive for χ constructed in § 4. We know - remark 4.2 - that the Hodge structures $H_\sigma(M(\chi))$, for $\sigma: K \hookrightarrow \overline{\mathbb{Q}} \subset \mathbb{C}$ are given by

$$H_\sigma(M(\chi)) \otimes_{E,\tau} \mathbb{C} \subset H^{n(\sigma,\tau),\, w-n(\sigma,\tau)} ,$$

the $n(\sigma,\tau)$ describing the infinity type of χ as in **0**, § 4. Every $H_\sigma(M(\chi))$ is a CM Hodge structure: in fact, the elements of E define endomorphisms of the motive $M(\chi)$, and therefore in particular elements of type (0,0) in $T^{1,1} = H_\sigma(M(\chi))^\vee \otimes H_\sigma(M(\chi))$, so that

$$MT(H_\sigma(M(\chi))) \subset GL_E(H_\sigma(M(\chi))/\mathbb{Q} = R_{E/\mathbb{Q}}\mathbb{G}_m;$$

on the other hand, we shall show that, for all σ, the corresponding map ν_σ factors through Z_L, for a suitable number field L. More precisely, remember that, when working with the Serre group Z, we consider all number fields as embedded into $\overline{\mathbb{Q}}$ (cf. **0**, 7.3.2). Pick a finite Galois extension L of $\mathbb{Q}$, $L \subset \overline{\mathbb{Q}}$, which contains K and E. Then

$$\nu_\sigma : R_{L/\mathbb{Q}}\mathbb{G}_m \to MT(H_\sigma(M(\chi))) \subset R_{E/\mathbb{Q}}\mathbb{G}_m$$

exists, and is given by the formula in the proof of 6.1.1 above. Now, $\mu^*(\sum_\tau n_\tau \cdot \tau) = \sum_\tau n(\sigma,\tau)n_\tau$, and thus one easily finds that ν_σ, as a homomorphism over $\mathbb{Q}$: $R_{L/\mathbb{Q}}\mathbb{G}_m \to R_{E/\mathbb{Q}}\mathbb{G}_m$, is given precisely by the array of numbers $\{n(s\sigma,\tau) \mid s \in G(L/\mathbb{Q}), \tau \in \mathrm{Hom}(E,\overline{\mathbb{Q}})\}$; cf. **0**, § 4. As the $n(\sigma',\tau)$'s come from the infinity-type of a Hecke character, we know - **0**, 7.3.2 - that ν_σ factorizes through Z_L, and therefore defines a representation $\tilde{\nu}_\sigma : Z \to R_{E/\mathbb{Q}}\mathbb{G}_m/\mathbb{Q}$ - which then corresponds to $H_\sigma(M(\chi))$ via 6.1.4. If σ is just the fixed embedding of K into $\overline{\mathbb{Q}}$, this representation ν_σ is nothing but the infinity type of χ, viewed as a representation of Z as in **0**, 7.3.2.

<u>6.1.6 Corollary</u> <u>Let</u> $\overline{\mathbb{Q}}$ <u>be the algebraic closure of</u> $\mathbb{Q}$ <u>in</u> $\mathbb{C}$, <u>and</u> σ <u>some embedding</u> $\overline{\mathbb{Q}} \hookrightarrow \mathbb{C}$. <u>The pair</u> $(CM_{\overline{\mathbb{Q}}}, H_\sigma)$ - <u>see</u> § 4 <u>above</u> - <u>is a neutralized tannakian category over</u> $\mathbb{Q}$ <u>with</u> Z <u>as corresponding group scheme.</u>

<u>Proof.</u> We show that H_σ establishes an equivalence of categories between the motives $CM_{\overline{\mathbb{Q}}}$ and all CM Hodge structures. By theorem 2.4.3 we are reduced to showing that

(a) if M is a motive in $CM_{\overline{\mathbb{Q}}}$, then $H_\sigma(M)$ is a CM Hodge structure;

(b) every CM Hodge structure arises in this way.

Since there are no non trivial Artin motives over $\overline{\mathbb{Q}}$, for (a) it suffices to show that $H_1^{\sigma}(A)$ is a CM Hodge structure if $A/\overline{\mathbb{Q}}$ is an abelian variety with complex multiplication as in 1.1 - but this is a special case of 6.1.5. As to (b), any representation of Z breaks up - over a suitable number field E - into a direct sum of characters $\lambda : Z \to R_{E/\mathbb{Q}}\mathbb{G}_m$. Any such λ is the infinity type of some Hecke character χ with values in E, defined over a suitable number field $K \subset \overline{\mathbb{Q}}$. We have seen in 6.1.5 how these infinity types arise from the Hodge structures $H_\sigma(M(\chi)) = H_\sigma(M(\chi) \times_K \overline{\mathbb{Q}})$.

6.1.7 Remark It follows (for example, from 6.1.4 and the definition of $CM_{\overline{\mathbb{Q}}}$ ...) that every CM Hodge structure is polarizable. - In fact, it is more customary - cf. [Lg], p. 215 f - to *define* a CM Hodge structure as a polarizable rational Hodge structure whose Mumford-Tate group is abelian.

6.2 Taniyama extensions

We now start setting up the formalism for determining the group scheme corresponding to $(CM_{\mathbb{Q}}, H_B)$. Proofs are essentially omitted.

6.2.0 Definition. A *Taniyama extension* is an exact sequence of affine group schemes over $\mathbb{Q}$

$$1 \to Z \overset{i}{\to} T \overset{\varphi}{\to} \mathrm{Gal}(\overline{\mathbb{Q}}/\mathbb{Q}) \to 1,$$

where Z is the Serre group, together with a homomorphism of topological groups $\varepsilon : \mathrm{Gal}(\overline{\mathbb{Q}}/\mathbb{Q}) \to T(\mathbf{A}^f)$ such that $\varphi_{\mathbf{A}^f} \circ \varepsilon = \mathrm{id}$.

Implicit in this definition is the requirement that the action of T on Z by conjugation - which, as Z is abelian, factors through $\mathrm{Gal}(\overline{\mathbb{Q}}/\mathbb{Q})$ - be the *second* galois action described in 6.1.2.

Any Taniyama extension may be written as inverse limit of sequences (with finite adelic splittings ε_E)

(6.2.1) $1 \to Z_E \to T_E \to \mathrm{Gal}\,(E^{ab}/\mathbb{Q}) \to 1$

over finite normal extensions E of $\mathbb{Q}$ contained in $\overline{\mathbb{Q}}$.

Given 6.2.1, choose any set theoretic splitting

$$a_E : \mathrm{Gal}\,(E^{ab}/\mathbb{Q}) \to T_E(E)$$

and define, for $s \in \mathrm{Gal}\,(E^{ab}/\mathbb{Q})$,

(6.2.2) $c_E(s) = \varepsilon_E(s)^{-1}\, a_E(s) \pmod{Z_E(E)}$,

a class in $Z_E(E_{\mathbb{A}^f})/Z_E(E)$ which is independent of the choice of a_E. Next, for λ any character $Z_E \to \mathbb{G}_m$ defined over $\overline{\mathbb{Q}}$ and $s \in \mathrm{Gal}\,(\overline{\mathbb{Q}}/\mathbb{Q})$, define the "cocycle":

(6.2.3) $c_E(s,\lambda) = \lambda(c_E(s)) \in E^*_{\mathbb{A}^f}/E^*$.

These finite idèle classes have the following properties, valid for all $s,t \in \mathrm{Gal}\,(E^{ab}/\mathbb{Q})$; $\lambda,\lambda' \in \mathrm{Hom}_{\overline{\mathbb{Q}}}\,(Z_E,\mathbb{G}_m)$; λ^s being defined as in 6.1.2.

<u>6.2.4 Lemma</u> (i) $c_E(st,\lambda) = c_E(s,\lambda^t)\, c_E(t,\lambda)$.

(ii) $c_E(t,\lambda)^s = c_E(t,\lambda^s)$.

(iii) <u>If</u> $F \supset E$, <u>then</u> $c_F(s,\lambda \circ N_{F/E}) = c_E(s,\lambda)$.

(iv) <u>If</u> c = complex conjugation, <u>then</u>: $c_E(c,\lambda) = 1$ <u>if and only if</u> $\varepsilon(c) \in T(\mathbb{Q})$.

(v) $c_E(s,\lambda)\, c_E(s,\lambda') = c_E(s,\lambda\cdot\lambda')$.

(In (iii), use that $E^*_{\mathbb{A}^f}/E^* \hookrightarrow F^*_{\mathbb{A}^f}/F^*$, by Hilbert 90.)

<u>6.2.5 Proposition. Two Taniyama extensions</u> T <u>and</u> T' <u>are isomorphic (as exact sequences of affine group schemes over</u> $\mathbb{Q}$ <u>with finite adelic splittings), if and only if the corresponding cocycles</u> c_E , c'_E <u>are equal, for all</u> E.

6.3 The group scheme for $(CM_{\mathbb{Q}}, H_B)$.

6.3.0 Let $\mathfrak{U}$ be the affine group scheme over $\mathbb{Q}$ which corresponds, by tannakian philosophy, to the neutralized category of motives $(CM_{\mathbb{Q}}, H_B)$, defined in § 4 above. Then $\mathfrak{U}$ is naturally endowed with the structure of a Taniyama extension

$$1 \to Z \xrightarrow{i} \mathfrak{U} \xrightarrow{\varphi} \mathrm{Gal}(\overline{\mathbb{Q}}/\mathbb{Q}) \to 1.$$

In fact, φ is given (via 2.3.2, (b)) by the fact that the Artin motives (2.4.1) are contained in $CM_{\mathbb{Q}}$ (note that $\mathfrak{U}$ is, in fact, proreductive: cf. 2.3.4); and i corresponds via 2.3.2 (c) to the functor $CM_{\mathbb{Q}} \to CM_{\overline{\mathbb{Q}}}$, $M \mapsto M \times_{\mathbb{Q}} \overline{\mathbb{Q}}$ (the condition of 2.3.2, (c) is satisfied because every object M of $CM_{\overline{\mathbb{Q}}}$ is defined over some number field K, and $R_{K/\mathbb{Q}} M \in CM_{\mathbb{Q}}$ contains M as a direct factor if viewed over $\overline{\mathbb{Q}}$), where we make use of 6.1.6. The exactness is then straightforward; the fact that the galois action on Z comes out right is slightly more subtle (cf. [DMOS], IV, B), or at least confusing ... - Finally, the finite adelic splitting ε required for a Taniyama extension, comes from the fact that the étale realization $H_{\mathbb{A}^f}(M)$ of a motive M of $CM_{\mathbb{Q}}$ carries an action of $\mathrm{Gal}(\overline{\mathbb{Q}}/\mathbb{Q})$, and is, on the other hand, isomorphic to $H_B(M) \otimes \mathbb{Q}_{\mathbb{A}^f}$.

6.3.1 We shall now write down, generalizing Tate (cf. [LCM], 7 § 3; [Bl] , § 4), a "cocycle" $g_E(s,\lambda)$, for every CM field E normal over $\mathbb{Q}$, which can be easily shown to be the cocycle corresponding to $\mathfrak{U}$ in the setup of 6.2. Let $M \in CM_{\overline{\mathbb{Q}}}(E)$ be of rank 1. For any $s \in \mathrm{Gal}(\overline{\mathbb{Q}}/\mathbb{Q})$, the conjugate motive M^s is defined in $CM_{\overline{\mathbb{Q}}}$ and carries the action of E transported by s, with respect to which it is also of rank 1 over E. Thus, fixing identifications

$$E \xrightarrow[\sim]{\theta} H_B(M)$$

$$E \xrightarrow[\sim]{\xi} H_B(M^s)$$

(where H_B denotes the realization H_σ, for σ the identical embedding $\overline{\mathbb{Q}} \hookrightarrow \mathbb{C}$), there is an element $a \in E^*_{\mathbb{A}^f}$ such that the following diagram commutes.

$$\begin{array}{ccc} E_{\mathbb{A}^f} & \xrightarrow{\theta \otimes \mathbb{Q}_{\mathbb{A}^f}} & H_{\mathbb{A}^f}(M) \\ \downarrow{\scriptstyle \cdot a} & & \downarrow{\scriptstyle s} \\ E_{\mathbb{A}^f} & \xrightarrow{\xi \otimes \mathbb{Q}_{\mathbb{A}^f}} & H_{\mathbb{A}^f}(M^s) \end{array}$$

6.3.2 Theorem *Up to multiplication by an element of* E^*, a *depends only on* s *and on the representation* $\lambda: Z \to R_{E/\mathbb{Q}}\mathbb{G}_m/\mathbb{Q}$ *corresponding to the CM Hodge structure* $H_B(M)$.

The proof is easy from what we already know - but it does, of course, use the absolute Hodge cycle theorem . - Rewriting $\lambda: Z \to R_{E/\mathbb{Q}}\mathbb{G}_m/\mathbb{Q}$ as $\lambda: Z_E \to \mathbb{G}_m$, defined over E, we shall write the class $a \cdot E^*$ as $g_E(s,\lambda)$, thereby defining the cocycle characterizing $\mathfrak{U}$.

6.3.3 Lemma (i) $g_E(s,\lambda)$ *has the properties analogous to* (i),(ii),(iii) and (v) of 6.2.4.

(ii) $g_E(c,\lambda) = 1$, *for all* λ.

(iii) $g_E(s,\lambda)^{1+c} = \Psi(s)^{-w} \cdot E^*$,

where $\Psi(s) \in \hat{\mathbb{Z}}^*$ *is such that* $\zeta^s = \zeta^{\Psi(s)}$, *for every root of unity* $\zeta \in \overline{\mathbb{Q}}^*$, *and* w *is the weight of the Hodge structure* $H_B(M)$.

In the case where λ is a CM type (i.e., $M = H_1(A)$, for some CM abelian variety $A/\overline{\mathbb{Q}}$) and $\lambda^s = \lambda$, the class $g_E(s,\lambda)$ is given by the Shimura-Taniyama reciprocity law. This class field theoretic description of $g_E(s,\lambda)$ will be generalized in 6.4, and the fact that it does describe g_E will be equivalent to the isomorphism between $\mathfrak{U}$ and the Taniyama group ...

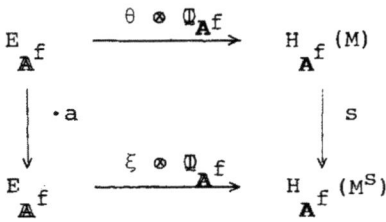

6.4 The Taniyama group

We proceed to define Tate's second cocycle $f_E(s,\lambda)$- generalizing it the same way we generalized g_E in 6.3: cf. [LCM], 7 §§ 1,2, and [Bl], § 4.

6.4.0 First, generalize Tate's "half transfer":

Given a CM field E normal over $\mathbb{Q}$, choose a system of representatives (remember that $E \subset \overline{\mathbb{Q}}$)

$$v:\mathrm{Hom}\,(E,\overline{\mathbb{Q}}) \to \mathrm{Gal}\,(E^{ab}/\mathbb{Q})$$

such that $v(\tau)_{|E} = \tau$ and $v(c\tau) = cv(\tau)$, for c = complex conjugation. Then, for $s \in \mathrm{Gal}(\overline{\mathbb{Q}}/\mathbb{Q})$ and $\lambda : Z_E \to \mathbb{G}_m$ over E (or over $\mathbb{Q}$), write λ as

$$\lambda = \sum_{\tau:E\hookrightarrow\overline{\mathbb{Q}}} n_\tau \, \tau \quad ,$$

and set

$$V_E(s,\lambda) = \prod_{\tau:E\hookrightarrow\overline{\mathbb{Q}}} (v(s\tau)^{-1}\cdot(s_{|E^{ab}})\cdot v(\tau))^{-n_\tau} \quad ,$$

an expression well defined in $\mathrm{Gal}(E^{ab}/\mathbb{Q})$.

6.4.1 Notations In the situation of 6.4.0, denote by $r_E : E^*_{\mathbb{A}f}/E^* \to \mathrm{Gal}(E^{ab}/E)$ the reciprocal of the classical Artin map, i.e., r_E sends a uniformizer π to a geometric Frobenius at π. - Recall the cyclotomic character Ψ defined in 6.3.3 (iii), and note that one has

$$r_{\mathbb{Q}}(\Psi(s)) = s_{|\mathbb{Q}^{ab}} \quad ,$$

for all $s \in \mathrm{Gal}(\overline{\mathbb{Q}}/\mathbb{Q})$. - Finally, for λ as before, write as usual $w = n_\tau + n_{c\tau}$ (any τ) the weight of λ (or: of the corresponding Hodge structure).

6.4.2 Proposition/Definition. With the preceding notations, there exists, for any $s \in Gal(\bar{\mathbb{Q}}/\mathbb{Q})$, a unique class $f_E(s,\lambda) \in E^*_{\mathbb{A}f}/E^*$, such that

(i) $r_E(f_E(s,\lambda)) = V_E(s,\lambda)$,

and (ii) $f_E(s,\lambda)^{1+c} = \Psi(s)^{-w} \cdot E^*$.

For the proof, cf. [LCM], 7, 2.2.

6.4.3 One can show that $f_E(s,\lambda)$ gives the cocycle attached to the Taniyama group $\mathfrak{T}$ defined by Langlands in [Lg], § 5 - except possibly for a certain number of renormalizations, of the sort carried out in [DMOS], III. We have not taken the time to check the details, and for us $\mathfrak{T}$ will be the Taniyama extension characterized by $f_E(s,\lambda)$ (whose existence is not proved here.) We do however call this $\mathfrak{T}$ the Taniyama group.

6.5 The Main Theorem, Consequences

6.5.1 Theorem. The cocycles $f_E(s,\lambda)$ and $g_E(s,\lambda)$ are equal: the Taniyama extension $\mathfrak{U}$ corresponding to $(CM_{\mathbb{Q}}, H_B)$ is isomorphic to the Taniyama group $\mathfrak{T}$.

The reader may obtain a complete proof of $f_E = g_E$ from [LCM], chap. 7 : the one crucial property of g (therefore of "e") not demonstrated in Lang - theorem 4.2 of [LCM], 7 - has simply been built into our motivic construction of $g_E(s,\lambda)$! In translating Lang's setup into our notations one has to identify a CM abelian variety $A/\mathbb{C}$ with the motive $H_1(A)$ - cf. 1.1 above. Thus instead of the CM types Φ (of weight + 1) in Lang, we consider representations λ with $n_\tau = -1$ or 0, of weight -1.

6.5.2 We shall use the following notation for the Taniyama group $\mathfrak{T}$:

$$\begin{array}{ccccccccc} 1 & \to & Z & \xrightarrow{j} & \mathfrak{T} & \xrightarrow{\phi} & \mathrm{Gal}\,(\overline{\mathbb{Q}}/\mathbb{Q}) & \to & 1 \\ & & & & & & \downarrow{\scriptstyle \alpha} & & \\ & & & & \mathfrak{T}(\mathbb{A}^f) & \leftarrow & & & \end{array}$$

We write ${}_K\mathfrak{T} = \phi^{-1}(\mathrm{Gal}\,(\overline{\mathbb{Q}}/K))$, for any number field $K \subset \overline{\mathbb{Q}}$. And if E is a finite normal extension of $\mathbb{Q}$, also contained in $\overline{\mathbb{Q}}$, such that $K \subset E^{ab}$, then we write ${}_K\mathfrak{T}_E$ for the image of ${}_K\mathfrak{T}$ in the quotient $\mathfrak{T}_E$ (6.2.1) of $\mathfrak{T}$. One might call

$$\begin{array}{ccccccccc} 1 & \to & Z & \to & {}_K\mathfrak{T} & \to & \mathrm{Gal}\,(\overline{\mathbb{Q}}/K) & \to & 1 \\ & & & & & & \downarrow{\scriptstyle \alpha|G(\overline{\mathbb{Q}}/K)} & & \\ & & & & {}_K\mathfrak{T}(\mathbb{A}^f) & \leftarrow & & & \end{array}$$

a Taniyama extension *over* K. It is the inverse limit of

(6.5.3) $$\begin{array}{ccccccccc} 1 & \to & Z_E & \to & {}_K\mathfrak{T}_E & \to & \mathrm{Gal}(E^{ab}/K) & \to & 1 \\ & & & & & & \downarrow{\scriptstyle \alpha_E} & & \\ & & & & {}_K\mathfrak{T}_E(\mathbb{A}^f) & \leftarrow & & & \end{array}$$

Note that ${}_K\mathfrak{T}_E$ is abelian, if $K \supset E$.

6.5.4 Scholion. Let $K \subset \overline{\mathbb{Q}}$ *be a number field, and write* H_B *for the fibre functor* H_σ *on* CM_K, *with* σ = *the inclusion* $K \hookrightarrow \overline{\mathbb{Q}}$. *Then* ${}_K\mathfrak{T}$ *is the affine group scheme corresponding to* (CM_K, H_B).

Cf. [DMOS], p. 265.

6.5.5 Scholion. Let $K \subset \overline{\mathbb{Q}}$ *be a number field which is galois over* $\mathbb{Q}$. *Then the structure* ${}_K\mathfrak{T}_K$ *(i.e., 6.5.3 with* $E = K$) *is isomorphic to Serre's group* S_K, *i.e. to the sequence* **0**, *7.3.1, equipped with the section* ε *of* **0**, *7.4. - Equivalently, there is an isomorphism*

$$ {}_K\mathfrak{T}^{ab} \cong S_K $$

<u>compatible with the finite adelic splittings</u> α <u>and</u> ε .

Two proofs of 6.5.5 are possible: First, a direct proof using only the cocycle f_E characterizing $\mathfrak{T}$ - cf. [Lg] , p. 224; second, using the fact that $\mathfrak{T} \cong \mathfrak{U}$, one can exploit the existence of $M(\chi)$ in CM_K, for any Hecke character χ of K, to identify $\mathrm{Hom}_{\overline{\mathbb{Q}}}({}_K\mathfrak{T}, \mathbb{G}_m)$ with $\mathrm{Hom}_{\overline{\mathbb{Q}}}(S_K, \mathbb{G}_m)$ - cf. [DMOS], IV, (D) and (E).

<u>6.5.6</u> The first proof of 6.5.5 would provide us with a <u>new method to construct</u> $M(\chi)$ (via 6.5.1): Viewed as a representation ${}_K\mathfrak{T} \to R_{E/\mathbb{Q}}\mathbb{G}_m$ over $\mathbb{Q}$, the motive $M(\chi)$ is simply the Hecke character χ of K with values in E, interpreted as in $\mathbb{D}$, 7.2,

$$ S_K \overset{\chi}{\to} R_{E/\mathbb{Q}}\mathbb{G}_m \quad \text{over } \mathbb{Q} , $$

and pulled back to ${}_K\mathfrak{T}$, via the canonical map ${}_K\mathfrak{T} \to {}_K\mathfrak{T}^{ab}$.

<u>6.5.7 Corollary.</u> <u>Let</u> K <u>and</u> E <u>be number fields, and</u> $K \subset \overline{\mathbb{Q}}$. <u>Let</u> M <u>be a motive in</u> $CM_K(E)$. <u>Then the system of</u> λ-<u>adic representations, for</u> λ <u>running over the finite places of</u> E,

$$ H_\lambda(M) = H_\ell(M) \otimes_{E \otimes \mathbb{Q}_\ell} E_\lambda \qquad (\text{where } \lambda | \ell) $$

<u>is a strictly compatible system of</u> E-<u>rational Galois representations.</u>

Recall that the statement of the corollary means that there is a finite set Σ of places of K such that for any prime ideal $\mathfrak{p}$ of K not contained in Σ , and any place λ of E with $\mathfrak{p} \nmid \mathbb{N}\lambda$, the Galois representation $H_\lambda(M)$ is unramified at $\mathfrak{p}$ (so that the action of a geometric Frobenius element $\mathrm{Frob}\,\mathfrak{p}$ at $\mathfrak{p}$ on $H_\lambda(M)$ is well defined), and the "characteristic polynomial"

$$\det_{E_\lambda}(1 - \mathrm{Frob}\,\mathfrak{p} \cdot X \mid H_\lambda(M)) \in E_\lambda[X]$$

actually has coefficients in $E \hookrightarrow E_\lambda$ which are independent of the place λ.

One proof of 6.5.7 (via 6.5.1) uses the fact - due to Langlands, [Lg], p. 226/227; cf. [DMOS], III, 3.17 - that there is a natural homomorphism $W_{\mathbb{Q}} \to \mathfrak{T}(\mathbb{C})$, where $W_{\mathbb{Q}}$ is the global Weil group of $\mathbb{Q}$. See [DMOS], IV, remarques 2,3 (p. 262). - As Greg Anderson has pointed out (see [A2], 5.7), 6.5.7 as well as a few other important corollaries of 6.5.1 can also be obtained using R. Brauer's induction lemma:

<u>6.5.8 Lemma</u> <u>The Grothendieck group of</u> $Rep_{\mathbb{C}}(\mathfrak{T} \times \mathbb{C})$ <u>is generated by the representations of the form</u>

$$\mathrm{Ind}_{K/\mathbb{Q}}(\chi),$$

<u>for number fields</u> K <u>and characters</u> $\chi : {}_K\mathfrak{T} \to \mathbb{G}_m/\mathbb{C}$.

In fact, $\mathfrak{T} \times \mathbb{C}$ is the inverse limit of $\mathbb{C}$-algebraic groups whose connected components are tori. And the Grothendieck group of $Rep_{\mathbb{C}}(G)$, for G a $\mathbb{C}$-algebraic group such that G^0 is a torus, is generated by the representations induced from 1 dimensional characters of subgroups of finite index.

Now, to deduce 6.5.7 from 6.5.8, assume (without loss of generality, applying $R_{K/\mathbb{Q}}$) that $K = \mathbb{Q}$ in the claim of 6.5.8, and note simply that

$$\det_E(1 - \mathrm{Frob}\,\mathfrak{p}\cdot X \mid \mathrm{Ind}_{K/\mathbb{Q}}(\chi)) = \prod_{\mathfrak{p}\mid p} (1 - \chi(\mathfrak{p})\cdot X) \in E[X],$$

at good places $\mathfrak{p}$.

<u>6.5.9</u> The corollary 6.5.7 allows to unconditionally define the <u>L-function of a motive</u> M <u>in</u> $CM_K(E)$ (or, equivalently, M in $CM_{\mathbb{Q}}(E)$): For $s \in \mathbb{C}$ with $\mathrm{Re}(s) >> 0$, put (cf. 1.8 above):

$$L^*(M/K,s) = \prod_{\mathfrak{p}} \det_E(1 - \mathrm{Frob}\,\mathfrak{p} \cdot \mathbb{N}\mathfrak{p}^{-s} \mid H_\lambda(M)^{I_{\mathfrak{p}}})^{-1} ,$$

where $\mathfrak{p}$ runs over all finite primes of K, and the determinant is calculated using any λ such that $\mathfrak{p} \nmid \mathbb{N}\lambda$. It is well known that the product converges for Re(s) sufficiently big, defining an element

$$L^*(M/K,s) \in E \otimes \mathbb{C} \cong \mathbb{C}^{\mathrm{Hom}(E,\mathbb{C})} .$$

(With our definition of strict compatibility, we have no control, a priori, about the Euler factors of the primes $\mathfrak{p}$ in the bad set Σ. This problem disappears however, in the light of 6.5.8!)

If $E = \mathbb{Q}$ (i.e., M is considered without E action), then we write simply L(M/K,s) for the ("Hasse-Weil") L-function of the motive M.

Recall that, as functions on $\mathbb{C}$ (a priori on $\{\mathrm{Re}(s) >> 0\}$), the following L-functions coincide:

$$L^*(M/K,s) = L^*(R_{K/\mathbb{Q}}M/\mathbb{Q},s).$$

(This generalizes the identity recalled before 6.5.9.)
In terms of L-functions, 6.5.8 reads:

<u>6.5.10 Corollary.</u> <u>For any motive</u> M <u>in</u> CM_K, <u>there exist number fields</u> $L_1,\dots,L_n$ <u>and, for every</u> $i = 1,\dots,n$, <u>an algebraic Hecke character</u> χ_i <u>of</u> L_i <u>with values in a suitable field</u> $E \subset \mathbb{C}$, <u>and an integer</u> m_i, <u>such that</u>

$$L(M/K,s) = \prod_{i=1}^{n} L(\chi_i,s)^{m_i} .$$

Here, $L(\chi_i, s)$ is the L-function $L(\chi_i^\tau, s)$ of **0**, § 6, with $\tau: E \hookrightarrow \mathbb{C}$.

6.5.11 Amazingly enough, the same line of thought also gives an alternative proof of the unicity theorem 5.1! For this, we refer to [A2] , 5.7.5.

6.6 Motives of rank 1 arising from abelian varieties

From 3.1, 5.3 and 6.5.7, we can now deduce that conjecture [DP] , 8.1, (iii) is also true - like all the rest of conjecture [DP], 8.1 - in the category M_K^{av} of motives over the number field K which can be constructed from the cohomology of abelian varieties (with or without complex multiplication). Joined with 5.1, this gives the

6.6.1 Theorem Every motive M in $\mathcal{M}_K^{av}(E)$, for a number field E, of rank 1 over E, is isomorphic in $\mathcal{M}_K^{av}$ to a motive $M(\chi)$ - see 4.1 - , for some algebraic Hecke character χ of K with values in E.

7. Anderson's motives for Jacobi sum characters

This section continues **0** § 8.

7.1 The basic example (Reference: [DMOS] , I § 7)

7.1.1 For integers $m \geq 0$, $n > 1$, let $X_m^n \overset{i}{\hookrightarrow} \mathbb{P}^{n-1}$ be the Fermat hypersurface of dimension $n - 2$ and degree m, given in projective coordinates by the equation

$$x_1^m + \ldots + x_n^m = 0.$$

The twisted primitive cohomology motive

$$h_{prim}(X_m^n)(-1) = [h(X_m^n)/i^*h(\mathbb{P}^{n-1})] \otimes \mathbb{Q}(-1),$$

a priori an object of $M_{\mathbb{Q}}$, decomposes over $\mathbb{Q}(\mu_m)$ under the action of the group

$$G_m^n = (\bigoplus_{i=1}^{n} \mu_m)/(\text{diagonal}) \subset \text{Aut}\,(X_m^n/\mathbb{Q}(\mu_m)).$$

Specifically, write the characters of G_m^n as

$$\underline{a} = \sum_{i=1}^{n} [a_i] \in \bigoplus_{i=1}^{n} \tfrac{1}{m}\mathbb{Z}/\mathbb{Z}, \quad \sum_{i=1}^{n} a_i = 0, \quad \text{all } a_i \neq 0,$$

with $\underline{a}((\zeta_1,\dots,\zeta_n)\,(\text{mod diag.})) = \prod_{i=1}^{n} \zeta_i^{a_i m}$.

Define the eigenmotive $h_{prim}(X_m^n)_{\underline{a}}$ as the image of $[h_{prim}(X_m^n) \otimes \mathbb{Q}(\mu_m)] \times \mathbb{Q}(\mu_m)$ under the projector

$$P_{\underline{a}} = \frac{1}{\# G_m^n} \sum_{g \in G_m^n} C(g) \otimes \underline{a}(g)^{-1},$$

where $C(g)$ is g, viewed as endomorphism (= absolute Hodge correspondence) of $h(X_m^n)$. Here, $\mathbb{Q}(\mu_m)$ in the tensor product is (the first component of) the unit object of $\mathcal{M}_{\mathbb{Q}}(\mathbb{Q}(\mu_m))$ - cf. 3.0 above. One shows that $h_{prim}(X_m^n)_{\underline{a}}$ is an object of $\mathcal{M}_{\mathbb{Q}(\mu_m)}(\mathbb{Q}(\mu_m))$ of rank 1, and that its L-function is given, in terms of the Jacobi sum Hecke characters of **0**, 8.2, by the relation [see 6.5.9 for the notation] :

$$L^*(h_{prim}(X_m^n)_{\underline{a}}(-1)/\mathbb{Q}(\mu_m),s) = (L(J_{\mathbb{Q}(\mu_m)}(c\underline{a})^{\tau},s))_{\tau \in \text{Hom}(\mathbb{Q}(\mu_m),\mathbb{C})}$$

7.1.2 As Jacobi sum Hecke characters are galois equivariant - see **0**, 8.2.5 - all the components of this array of L-functions are actually equal (as meromorphic functions on $\mathbb{C}$). On the other hand, the sum of projectors

$$\bigoplus_{\sigma \in \text{Gal}(\mathbb{Q}(\mu_m)/\mathbb{Q})} P_{\sigma\underline{a}}$$

(the galois action being that of **0**.8.2.1) is an absolute Hodge correspondence of $h(X_m^n) \otimes \mathbb{Q}(\mu_m)$ defined over $\mathbb{Q}$. Thus, there is a motive $M(\underline{a})$ in $\mathcal{M}_{\mathbb{Q}}$, with coefficients in $\mathbb{Q}$ (the action of G_m^n is only defined over $\mathbb{Q}(\mu_m)$) of rank $[\mathbb{Q}(\mu_m):\mathbb{Q}]$ such that

$$M(\underline{a})(1) \times \mathbb{Q}(\mu_m) \cong h_{prim}(X_m^n)_{c\underline{a}}$$

in $M_{\mathbb{Q}(\mu_m)}$; consequently

$$L(M(\underline{a})/\mathbb{Q},s) = L(J_{\mathbb{Q}(\mu_m)}(\underline{a}),s)$$

(an identity of functions on $\mathbb{C}$.)

7.1.3 The motive $h(X_m^n)$, and therefore also $M(\underline{a})$, is (isomorphic to) a motive in $M_{\mathbb{C}}^{av}$, and in fact, by the same token, in $CM_{\mathbb{Q}}$ - cf. 5.3.

This is shown by Shioda induction: see [A2], § 9 for a detailed proof of how to express $h(X_m^n)$ in terms of the Fermat curves X_m^2, X_m^3 and of $\mathbb{P}^1$. As the Jacobian of Fermat curves are well known to admit complex multiplication (over $\mathbb{Q}(\mu_m)$) this directly proves the stronger assertion.

7.1.4 Thus we have indicated how to construct, for $\underline{a}$ as above, a motive $M(\underline{a})$ in $M_{\mathbb{Q}}^{av}$, and in fact a representation of the Taniyama group, whose L-function is the Hecke L-function of $J_{\mathbb{Q}(\mu_m)}(\underline{a})$. Note that, in view of 6.5.7, this already proves that $J_{\mathbb{Q}(\mu_m)}(\underline{a})$ is a Hecke character, and more precisely, a Hecke character unramified outside m - because this is true of the ℓ-adic representations of X_m^n.
Anderson, in [A1], and especially in [A2], has generalized this construction of $M(\underline{a})$ to all Jacobi sum Hecke characters, in the sense of **0**, 8.2.4.

7.2 Anderson's first theorem (Reference: [A1] or [A2])

Let K be an abelian number field and $\underline{a} \in \mathbb{B}_K^0$ (for the notation, see **0**, 8.2). There is a motive $M_K(\underline{a})$ in $CM_{\mathbb{Q}}$ of rank $[K:\mathbb{Q}]$ such that

$$L(M_K(\underline{a}),s) = L(J_K(\underline{a}),s).$$

Upon extension of scalars, $M_K(\underline{a}) \times K$ acquires the structure of a motive of rank 1 in $CM_K(K)$, which is a motive for $J_K(\underline{a})$, in the sense of 3.3.

The crucial point about this theorem is that $M_K(\underline{a})$ is constructed from the cohomology of Fermat hypersurfaces X_m^n. This will make it possible to calculate the periods of $M_K(\underline{a})$ in terms of values of the Γ-function at rational numbers: see II § 4! In [A1], $M_K(\underline{a})$ is explicitly constructed as sitting in twisted Fermat hypersurfaces; in [A2], the theorem is no longer stated the way we just did but rather embedded in a much more general formalism which we shall now sketch very roughly.

7.3 Anderson's ulterior motives (Reference: [A2])

7.3.1 An arithmetic Hodge structure W of weight $w \in \mathbb{Z}$ is

- a finite dimensional $\mathbb{Q}$ vector space W_B, with a decomposition

$$W_B \otimes \mathbb{C} = \bigoplus_{\substack{a,b\in\mathbb{Q}\\ a+b=w}} W^{a,b}$$

such that $(1 \otimes c)W^{a,b} = W^{b,a}$, for c = complex conjugation;

- a $\mathbb{Q}$-linear subspace W_{DR} of $W_B \otimes \mathbb{C}$ of the same dimension as W_B such that

(i) for all $a \in \mathbb{Q}/\mathbb{Z}$, writing

$$F^a(W_B \otimes \mathbb{C}) = \bigoplus_{\substack{a'\in\mathbb{Q}\\ a'\geq a}} W^{a',w-a'} \quad \text{and} \quad F^a W_{DR} = W_{DR} \cap F^a(W_B \otimes \mathbb{C}),$$

one has

$$(F^a W_{DR}) \otimes \mathbb{C} = F^a(W_B \otimes \mathbb{C});$$

(ii) there is a $\mathbb{Q}$-linear involution $F_\infty : W_B \to W_B$ making the following triangle commute.

$$\begin{array}{ccc} W_B \otimes \mathbb{C} & \xrightarrow{\;F_\infty \otimes c\;} & W_B \otimes \mathbb{C} \\ & \nwarrow \quad \nearrow & \\ & W_{DR} & \end{array}$$

An <u>arithmetic Hodge structure</u> is a finite direct sum of arithmetic Hodge structures of fixed weights.
It is obvious how the notion of arithmetic Hodge structure is a generalization of the "part at infinity" of a motive over $\mathbb{Q}$ - cf. [DP], 1.4 , for the last requirement, (ii).
We shall write $\omega_\infty(M)$ for the arithmetic Hodge structure given by a motive M in $\mathcal{M}_{\mathbb{Q}}$. The fractional exponents permitted in the decomposition of $W_B \otimes \mathbb{C}$ will be needed to accomodate individual Gauss sums ...

A morphism of arithmetic Hodge structures is a $\mathbb{Q}$-linear map of the W_B's which respects the $W^{a,b}$'s and the W_{DR}'s.
Just like Hodge structures, arithmetic Hodge structures form a tannakian category over $\mathbb{Q}$, say AH , neutralized by the functor $W \mapsto W_B$ into $\mathbb{Q}$ vector spaces.

<u>7.3.2</u> Write $2\pi i\, \hat{\mathbb{Z}}$ for the Pontrjagin dual of $\mathbb{Q}/\mathbb{Z}$, and consider the pairing

$$\begin{array}{ccc} 2\pi i\, \hat{\mathbb{Z}} \times \mathbb{Q}/\mathbb{Z} & \longrightarrow & \mathbb{C}^* \\ (2\pi i n,\ a) & \longmapsto & \langle 2\pi i n, a\rangle = \exp(2\pi i \langle na\rangle), \end{array}$$

where the function $\langle . \rangle$ on $\mathbb{Q}/\mathbb{Z}$ was defined in **0**, 8.1.4.
Given $0 \neq a \in \mathbb{Q}/\mathbb{Z}$, define $\gamma(a): 2\pi i\, \hat{\mathbb{Z}} \to \mathbb{C}^*$ by

$$\gamma(a)(2\pi i\, n) = \langle 2\pi i\, n, a\rangle \,.\, \Gamma(\langle -a\rangle).$$

For each integer $m \geq 1$, we define the arithmetic Hodge structure of weight 1, E_m by:

- $(E_m)_B = \left\{ e: 2\pi i\, \hat{\mathbb{Z}} \to \mathbb{Q} \;\middle|\; \begin{array}{l} e \text{ factors through } 2\pi i(\hat{\mathbb{Z}}/m\hat{\mathbb{Z}}) \\ \text{and } \sum\limits_{j \in \frac{1}{m}\mathbb{Z}/\mathbb{Z}} e(2\pi i m j) = 0 \end{array} \right\}$

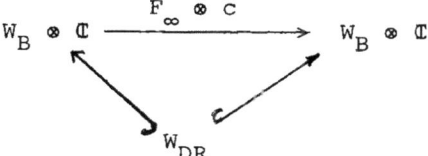

- $E_m^{\langle a\rangle,\langle -a\rangle} = \mathbb{C}\cdot\gamma(a)$, if $0 \neq a \in \frac{1}{m}\mathbb{Z}/\mathbb{Z}$

- $(E_m)_{DR} = \sum_{0\neq a\in\frac{1}{m}\mathbb{Z}/\mathbb{Z}} \mathbb{Q}\cdot\gamma(a)$.

7.3.3 Call $\widetilde{CM}$ the smallest tannakian subcategory of AH containing $\omega_\infty(M)$, for all objects M of $CM_{\mathbb{Q}}$, and E_m, for all $m \geq 1$. Write $\widetilde{\mathfrak{T}}$ for the affine group scheme over $\mathbb{Q}$ which corresponds to $(\widetilde{CM}, W \mapsto W_B)$. The $\mathbb{Q}$ vector space $(E_m)_B$, viewed as a representation of $\widetilde{\mathfrak{T}}$, is denoted $\mathbb{E}_m$ by Anderson, and he defines

$$\mathbb{E} = \varinjlim \mathbb{E}_m ,$$

using the inclusions of arithmetic Hodge structures $E_m \to E_n$, for $m|n$.

By tannakian philosophy, the functor $\omega_\infty : CM_{\mathbb{Q}} \to \widetilde{CM}$ corresponds to a morphism

$$\tilde{\phi} : \widetilde{\mathfrak{T}} \to \mathfrak{T}$$

with $\mathfrak{T}$ the Taniyama group (6.5).

Furthermore, there is an arrow

$$\tilde{\jmath} : 2\pi i\, \hat{\mathbb{Z}} \to \widetilde{\mathfrak{T}}$$

which arises as follows.

7.3.4 Let V be a $\mathbb{Q}$ vector space with an admissible $\mathbb{Q}$-linear action of $2\pi i\, \hat{\mathbb{Z}}$. Then V can be decomposed into eigenspaces over $\mathbb{C}$:

$$V \otimes \mathbb{C} = \bigoplus_{a\in\mathbb{Q}/\mathbb{Z}} V(a) ,$$

and one finds, for all $s \in \mathrm{Aut}\,\mathbb{C}$, that

$$(1\otimes s)\ V(a) = V(\Psi(s|_{\overline{\mathbb{Q}}})a)$$

with Ψ the cyclotomic character: see ◐, 8.2.1; or 6.4.1. Conversely, every decomposition respecting the galois action on $\mathbb{Q}/\mathbb{Z}$ comes from an (admissible) representation $2\pi i\ \hat{\mathbb{Z}} \to GL(V)$.

Now, given a representation W of $\tilde{\mathfrak{T}}$, and $a \in \mathbb{Q}/\mathbb{Z}$, put

$$W(a) = \bigoplus_{p,q\in\mathbb{Z}} W^{p+\langle a\rangle,\, q-\langle a\rangle}$$

The decomposition $W \otimes \mathbb{C} = \oplus\, W(a)$ is compatible with the galois action on $\mathbb{Q}/\mathbb{Z}$ - look at E_m ! - , and therefore comes from a representation $2\pi i\ \hat{\mathbb{Z}} \to GL(W)$. This action of $2\pi i\ \hat{\mathbb{Z}}$ depends naturally on W and thus defines the desired morphism $\tilde{\mathfrak{j}}: 2\pi i\ \hat{\mathbb{Z}} \to \tilde{\mathfrak{T}}$. Since motives have honest regard Hodge structures: with integral exponents, it is plain that the image of $\tilde{\mathfrak{j}}$ is contained in the kernel of $\tilde{\phi}$. - And more is true:

7.4 Anderson's second theorem ([A2], Theorem 8)

7.4.1 The sequence

$$1 \to 2\pi i\ \hat{\mathbb{Z}} \xrightarrow{\tilde{\mathfrak{j}}} \tilde{\mathfrak{T}} \xrightarrow{\tilde{\phi}} \mathfrak{T} \to 1$$

is exact.

7.4.2 Recall from 7.1.1 the motive $h_{prim}(X_m^n)(-1)$. According to 7.1.3, it may be viewed as giving a representation of the Taniyama group $\mathfrak{T}$, and hence, via $\tilde{\phi}$, a representation of $\tilde{\mathfrak{T}}$.

There is an isomorphism of $\tilde{\mathfrak{T}}$ representations

$$h_{prim}(X_m^n)(-1) \cong (\mathbb{E}_m^{\otimes n})^{2\pi i\ \hat{\mathbb{Z}}} .$$

The superscript denotes, of course, the subspace of elements invariant under $\tilde{\mathfrak{j}}(2\pi i\ \hat{\mathbb{Z}})$.

7.4.3 Use the embedding $\tilde{j}$, like in 7.3.4, to decompose

$$\mathbb{E} \otimes \mathbb{C} = \bigoplus_{a \in \mathbb{Q}/\mathbb{Z}} \mathbb{E}(a) .$$

Note that $\dim_{\mathbb{C}} \mathbb{E}(a) = 1$ or 0 according as $a \neq 0$ or $a = 0$ in $\mathbb{Q}/\mathbb{Z}$. Recall from 6.5.2 our notations for the Taniyama group.

For all $\tilde{t} \in \tilde{\mathfrak{T}}(\mathbb{C})$, calling $s \in \mathrm{Gal}(\overline{\mathbb{Q}}/\mathbb{Q})$ the image $\phi \circ \tilde{\phi}(\tilde{t})$, and for all $0 \neq a \in \mathbb{Q}/\mathbb{Z}$, one has

$$\tilde{t}\, \mathbb{E}(a) = \mathbb{E}(\Psi(s)^{-1}a).$$

7.4.4 Last but not least Anderson proves that there is a substitute for the galois action on the eigenspaces $\mathbb{E}(a)$, which shows their relation to Gauss sums. In stating it he makes use of a fixed chosen embedding of $\mathbb{Q}_\ell$ into $\mathbb{C}$, for every ℓ. In fact, recall $(\mathbb{O}, 8.2.2)$ that in the treatment of Jacobi sum Hecke characters we also fixed, at least, an extension of the absolute value $| \, |_\ell$ from $\mathbb{Q}$ to $\overline{\mathbb{Q}}$, for every ℓ.

Let p and ℓ be two distinct rational primes, and write $\mathrm{Frob}\, p \in \mathrm{Gal}(\overline{\mathbb{Q}}/\mathbb{Q})$ for a geometric Frobenius element (well determined up to $\mathrm{Gal}(\overline{\mathbb{Q}}/\mathbb{Q}^{ab})$ and up to the inertia group of the chosen extension of $| \, |_p$.) Call α_ℓ the ℓ-component of the splitting α of $\phi_{\mathbb{A}^f}$ - see 6.5.2.

There exists $\tilde{F}(p,\ell) \in \tilde{\mathfrak{T}}(\mathbb{C})$ satisfying

- $\alpha_\ell(\mathrm{Frob}\, p) = \tilde{\phi}(\tilde{F}(p,\ell))$
- for all positive integers f and all $0 \neq a \in \mathbb{Q}/\mathbb{Z}$ such that $(p^f-1)a = 0$ in $\mathbb{Q}/\mathbb{Z}$, one has

  $$\mathrm{tr}_{\mathbb{C}}(\tilde{F}(p,\ell)^f \,|\, \mathbb{E}(a)) = g_p\Big(\sum_{i=1}^{f} [p^i a]\Big),$$

  with g_p as in $\mathbb{O}$, 8.2.2.

<u>7.4.5</u> In the setup of Anderson's second theorem, the motives $M_K(\underline{a})$ of 7.2 are obtained like this: First, it is enough to consider elements $\underline{a} = \sum_{0 \neq a \in \mathbb{Q}/\mathbb{Z}} n_a[a] \in \mathbb{B}_K^0$ with $n_a \geq 0$ for all a. For such $\underline{a}$, put

$$\mathbb{E}(\underline{a}) = \bigotimes_a \mathbb{E}(a)^{\otimes n_a} .$$

$M_K(\underline{a})$ then appears as a $\mathbb{Q}$-rational representation of $\tilde{\mathfrak{T}}$ such that

$$M_K(\underline{a}) \otimes \mathbb{C} = \bigoplus_{\sigma \in \mathrm{Gal}(K/\mathbb{Q})} \mathbb{E}(\sigma\underline{a}) .$$

As $\underline{a} \in \mathbb{B}^0$, the action of $2\pi i\, \hat{\mathbb{Z}}$ on $M_K(\underline{a})$ is trivial which, by 7.4.1, makes it a representation of $\mathfrak{T}$, i.e., a motive in $CM_{\mathbb{Q}}$.

For all details of Anderson's construction we refer to [A2]. Our discussion of this work will be taken up again in II § 4 where we give an account of his period calculations.

<u>7.5</u> <u>Elliptic curves</u>

Let us mention in passing the geometric reasons that have motivated our choices of the basic characters of the exceptional imaginary quadratic fields: 0, 8.3.2.

It is easily checked that J_3 is the Hecke character of the elliptic curve

$$u^3 + v^3 = 1, \text{ or of } y(1-y) = x^3,$$

the latter one being $\mathbb{Q}$-isogenous to the first one. In fact, $u^3 + v^3 = 1$ is also the strong Weil curve for $\Gamma_0(27)$. These coincidences seemed to give some geometric privilege to this choice of J_3.

On a more historical basis, J_4 was chosen because it is the Hecke character of the famous elliptic curve $y^2 = x^3 - x$

which corresponds to the elliptic integral measuring the arc length of the lemniscate - cf. [HS], §1.

As to $\mathbb{Q}(\sqrt{-2})$, any Jacobi-sum Hecke character of infinity type 1 corresponds to a $\mathbb{Q}$-curve, in the sense of [Gr 1], §11. But we do not know of any such curve that has attracted individual interest.

CHAPTER TWO:

The Periods of Algebraic Hecke Characters

Although we introduce in this chapter the basic notion of our work, the constructions we have to present are quite formal. More precisely, we review what may be called the "arithmetic linear algebra" needed for our purposes:

- In § 1, we define the periods of a motive M in $\mathcal{M}_K(E)$ - K and E number fields -, and the periods $c^{\pm}(M)$, for M in $\mathcal{M}_{\mathbb{Q}}(E)$, introduced by Deligne [DP] to formulate his rationality conjecture for critical values of L-functions of motives. The periods of a Hecke Character χ of K with values in E are simply those of any motive M attached to χ in the sense of I.3.3, or the $c^{\pm}(R_{K/\mathbb{Q}}M)$.
- Deligne's rationality conjecture (and its proof in the case of Hecke characters) is recalled (resp. quoted) in § 2.
- § 3 is devoted to the study of the behaviour of our periods "under twisting". Very similar calculations have also been done by Blasius.
- § 4 continues and closes our discussion of Anderson's motives for Jacobi sum Hecke characters by recalling their periods. They will be needed in chapters III and IV.

1. The periods of a motive

1.0 Let K and E be finite extensions of $\mathbb{Q}$, and let M be a motive defined over K with coefficients in E, of rank r over E. Thus, in the notation of I.3.0, M is an object of $\mathcal{M}_K(E)$. But the linear algebra which we are about to present would work in any sensible theory of motives, not just the particular one using absolute Hodge cycles with which we work here. The constructions of this section are all (essentially) present in various sections of [DP].

1.1 Definition of p(M)

The component at infinity of the comparison isomorphism I,2.1.1 yields an isomorphism of free $K \otimes E \otimes \mathbb{C}$ modules of rank r,

$$(1.1.1) \qquad I : \bigoplus_{\sigma} H_{\sigma}(M) \otimes \mathbb{C} \xrightarrow{\sim} H_{DR}(M) \otimes \mathbb{C} ,$$

where an unmarked $\otimes$ is always over $\mathbb{Q}$ and σ runs through all distinct embeddings of K into $\mathbb{C}$; so the K-linear structure on the left hand side is obtained from identifying $K \otimes \mathbb{C}$ with $\mathbb{C}^{\mathrm{Hom}(K,\mathbb{C})}$.

The $K \otimes E \otimes \mathbb{C}$ modules compared in 1.1.1 are extensions of scalars up to $\mathbb{C}$ of

- the $E^{\mathrm{Hom}(K,\mathbb{C})}$ module $\bigoplus_{\sigma} H_{\sigma}(M)$, on the left;
- the $K \otimes E$ module $H_{DR}(M)$, on the right.

Choose bases $\gamma_1,\ldots,\gamma_r$ - where $\gamma_i = (\gamma_{i\sigma})_{\sigma}$ with $H_{\sigma}(M) = \bigoplus_{i=1}^{r} E\cdot\gamma_{i\sigma}$, for all σ -, resp. $\omega_1,\ldots,\omega_r$, of these sub - structures, and <u>define</u>

$$p(M) \in (K \otimes E \otimes \mathbb{C})^* = (\mathbb{C}^*)^{\mathrm{Hom}(K,\mathbb{C}) \times \mathrm{Hom}(E,\mathbb{C})}$$

to be the determinant of the matrix representing the isomorphism 1.1.1 relative to these bases.

Changing the bases multiplies $p(M)$ by an element of $(E^*)^{\mathrm{Hom}(K,\mathbb{C})}$, resp. of $(K \otimes E)^*$. Thus, $p(M)$ will be regarded modulo these operations, defining a class

$$p(M) \in (K \otimes E \otimes \mathbb{C})^*/(E^*)^{\mathrm{Hom}(K,\mathbb{C})} \cdot (K \otimes E)^* .$$

1.1.2 Remark. Let $\det_E M$ be the maximal E-linear exterior power of M , i.e., the direct factor of $M^{\otimes_E r}$ in $\mathcal{M}_K(E)$ whose σ-realization (for any $\sigma : K \hookrightarrow \mathbb{C}$) is $\bigwedge_E^r H_{\sigma}(M)$. Then $\det_E M$ is a motive of rank 1 over E , and it is clear from the definition of p that

(1.1.3) $p(\det_E M) = p(M)$.

1.2 Components of p(M)

Although p(M) is really a class $\mathrm{mod}(E^*)^{\mathrm{Hom}(K,\mathbb{C})}$ $(K \otimes E)^*$, we continue to think of our period, via representatives, as an array of complex numbers, in $(\mathbb{C}^*)^{\mathrm{Hom}(K,\mathbb{C}) \times \mathrm{Hom}(E,\mathbb{C})}$, and we write

(1.2.1) $p(M) = (p(M;\sigma,\tau))_{\sigma,\tau}$,

with $p(M;\sigma,\tau) \in \mathbb{C}^*$, for all $\sigma \in \mathrm{Hom}(K,\mathbb{C})$, $\tau \in \mathrm{Hom}(E,\mathbb{C})$.

Similarly, we occasionally write, for $\sigma : K \hookrightarrow \mathbb{C}$,

$$p(M;\sigma) = (p(M;\sigma,\tau))_\tau \in (E \otimes \mathbb{C})^* ,$$

and for $\tau : E \hookrightarrow \mathbb{C}$,

$$p(M;,\tau) = (p(M;\sigma,\tau))_\sigma \in (K \otimes \mathbb{C})^* .$$

(If K=E, the two roles of this field still have to be neatly separated!)

All these components are actually determinants (with respect to certain bases) of comparison isomorphisms derived from 1.1.1. For example, decompose our basis elements ω_i as

(1.2.2) $\omega_i = (\omega_{i\sigma})_\sigma \in H_{DR}(M) \otimes \mathbb{C} = \bigoplus_\sigma H_{DR}(M) \otimes_{K,\sigma} \mathbb{C}$.

By its very construction, 1.1.1 is the direct sum of isomorphisms

(1.2.3) $I_\sigma : H_\sigma(M) \otimes \mathbb{C} \xrightarrow{\sim} H_{DR} \otimes_{K,\sigma} \mathbb{C}$.

So, $p(M;\sigma)$ is the determinant of I_σ with respect to the bases $\{\gamma_{i\sigma}\}_i$ and $\{\omega_{i\sigma}\}_i$. - Note that $\{\omega_{i\sigma}\}_i$ is a $K^\sigma \otimes E$ basis of $H_{DR}(M) \otimes_{K,\sigma} K^\sigma$.

1.3 Field of coefficients

1.3.1 If $E' \supset E$ is a (finite) extension, and $M' = M \otimes_E E'$ - see I,3.0 -, then $H_\sigma(M') = H_\sigma(M) \otimes_E E'$, for all σ , and

$H_{DR}(M') = H_{DR}(M) \otimes_E E'$. Hence, $p(M')$ is simply the image of $p(M)$ under the natural map

$$(K \otimes E \otimes \mathbb{C})^* \hookrightarrow (K \otimes E' \otimes \mathbb{C})^* .$$

Observe that, if we know that M' in $\mathcal{M}_K(E')$ is of the form $M \otimes_E E'$, for some M in $\mathcal{M}_K(E)$, then $p(M)$ can be recuperated from $p(M')$ because, assuming E'/E to be galois with group G and letting G act trivially on K , one has

$$[(K\otimes E'\otimes \mathbb{C})^*/(E'^*)^{\mathrm{Hom}(K,\mathbb{C})}(K\otimes E')^*]^G = (K\otimes E\otimes \mathbb{C})^*/(E^*)^{\mathrm{Hom}(K,\mathbb{C})}(K\otimes E)^*!$$

In fact, use the exact sequence

$$1 \to E'^* \to (E'^*)^{\mathrm{Hom}(K,\mathbb{C})}(K\otimes E)^* \to \frac{(E'^*)^{\mathrm{Hom}(K,\mathbb{C})}(K\otimes E')^*}{(E'^*)^{\mathrm{Hom}(K,\mathbb{C})}\cap(K\otimes E')^*} \to 1 ,$$

as well as "Hilbert 90" for E'^* and $(K \otimes E')^*$ to conclude that

$$H^1(G,(E'^*)^{\mathrm{Hom}(K,\mathbb{C})}(K \otimes E')^*) = 0 .$$

1.3.2 Suppose now that M' in $\mathcal{M}_K(E')$ is given, and that $M = M'|_E$, in the notation of I,3.0. Using a basis $\{e_t'\}$ of E' over E to obtain bases $\{\gamma_i\}$ and $\{\omega_i\}$ for M from those chosen for M' , one finds that

$$p(M) = N_{E'/E}(p(M')) .$$

In terms of components, this means that

$$p(M;,\tau)= \prod_{\tau'|_E=\tau} p(M';,\tau'),$$

for $\tau \in \mathrm{Hom}(E,\mathbb{C})$ and $\tau' \in \mathrm{Hom}(E'\mathbb{C})$ restricting to τ .

1.4 Field of definition

1.4.1 If $K' \supset K$ is a finite extension and M is defined over K , then $p(M \times_K K')$ is clearly the image in $(K'\otimes E\otimes \mathbb{C})^*/(E^*)^{\mathrm{Hom}(K',\mathbb{C})}(K' \otimes E)^*$ of $p(M)$ via the natural map

$$K \otimes E \otimes \mathbb{C} \hookrightarrow K' \otimes E \otimes \mathbb{C} .$$

In practice it is often convenient to extend the base field in order to have eigendifferentials for the action of E defined over K' - cf. for example, 1.5.2 below. However, unlike 1.3.1, extending K does in general throw away information about $p(M)$: If K'/K is galois with group G acting trivially on E, then

$$H^1(G,(E^*)^{\mathrm{Hom}(K'\mathbb{C})}\cdot (K'\otimes E)^*) \cong \mathrm{Hom}(G,E^*) .$$

Put another way, different K'/K-forms of M' will in general have different periods $p(M)$.

1.4.2 Suppose now that M' in $\mathcal{M}_{K'}(E)$ is given, and define $M = R_{K'/K}M'$, in $\mathcal{M}_K(E)$. Then $H_{DR}(M)$ is $H_{DR}(M')$, but considered as $K \otimes E$ module. Therefore, for every $\sigma : K \hookrightarrow \mathbb{C}$,

$$H_{DR}(M) \otimes_{K,\sigma} \mathbb{C} = \bigoplus_{\sigma'|_K=\sigma} H_{DR}(M') \otimes_{K',\sigma'} \mathbb{C} ,$$

where σ' varies over the embeddings of K' that restrict to σ on K . Thus, if $\omega_i' = (\omega_{i\sigma'}')_{\sigma'}$ - like in 1.2.2 - make up a basis of $H_{DR}(M')$ over $K' \otimes E$, then $\{\omega_{i\sigma'}' \mid i=1,\dots,r';\sigma'|_K=\sigma\}$ is a basis of $H_{DR}(M) \otimes_{K,\sigma} \mathbb{C}$ over $E \otimes K \otimes_{K,\sigma} \mathbb{C}$. Here, r' is the rank of M' over E . And since

$$H_\sigma(M) = \bigoplus_{\sigma'|_K=\sigma} H_{\sigma'}(M') ,$$

it follows that

$$(1.4.3) \qquad p(M;\sigma) = \Big(\prod_{\sigma'|_K=\sigma} p(M';\sigma')\Big)\, D_\sigma ,$$

where $D_\sigma \in (E \otimes \mathbb{C})^*$ will now be computed. This factor comes in because the $\{\omega_{i\sigma'}'\}$ are not necessarily a basis of $H_{DR}(M) \otimes_{K,\sigma} K^\sigma$.

First, given the basis $\{\omega_i' \mid i=1,\dots,r'\}$ of $H_{DR}(M')$ over $K' \otimes E$, and choosing a basis $\{\alpha_s\}$ of K' over K , take

$\{\alpha_s \cdot \omega_i^!\}_{i,s}$ as basis of $H_{DR}(M)$ over $K \otimes E$. For every $\sigma : K \hookrightarrow \mathbb{C}$, the factor D_σ is then the determinant of $id_{H_{DR}(M)} \otimes_{K,\sigma} \mathbb{C}$, relative to the bases $\{\omega^!_{i\sigma'}\}_{i,\sigma'|_K=\sigma}$, on the left, and $\{(\alpha_s\omega_i^!)_\sigma\}_{i,s}$, on the right. Thus,

(1.4.4) $$D_\sigma = \det((\alpha_s^{\sigma'})_{s,\sigma'|_K=\sigma})^{-r'} \in \mathbb{C}^* \hookrightarrow (E \otimes \mathbb{C})^* .$$

Second, note that the undeterminacy of the determinants

(1.4.5) $$\delta(K'/K,\sigma) = \det((\alpha_s^{\sigma'})_{s,\sigma'|_K=\sigma})$$

is just what is allowed for in the definition of $p(M)$:
On the one hand, changing the basis $\{\alpha_s\}$ of K'/K multiplies $\delta(K'/K,\sigma)$ by k^σ, for some element $k \in K^*$, so that the array

(1.4.6) $$\delta(K'/K) = (\delta(K'/K,\sigma))_\sigma \in (K \otimes \mathbb{C})^* \hookrightarrow (K\otimes E\otimes \mathbb{C})^*$$

gets multiplied by $k \in K^* \hookrightarrow (K \otimes E)^*$. On the other hand, 1.4.5 gives $\delta(K'/K,\sigma)$ only up to a sign since no ordering was imposed on $\{\sigma'|_K = \sigma\}$, and in general, there does not seem to be a reasonable way to fix these signs simultaneously in σ. We are saved by the fact that $p(M)$ is only well defined up to factors in $(E^*)^{Hom(K,\mathbb{C})}$!

It is plain that $\delta(K'/K)^2 \in K^* \hookrightarrow (K \otimes E \otimes \mathbb{C})^*$. Thus, calling $\epsilon(M') \in \{0,1\}$ the rank of M' over E taken modulo 2 we have shown:

(1.4.7) $$p(R_{K'/K}M') = N_{K'/K}(p(M'))\ \delta(K'/K)^{\epsilon(M')} .$$

1.4.8 Remark. Recall the following characterizations of $\delta(K'/K)$, in the case that K' is normal over K. For $\sigma' : K' \hookrightarrow \mathbb{C}$ call σ the restriction of σ' to K. Then $K^\sigma(\delta(K'/K,\sigma)) \subseteq K'^{\sigma'}$ is the (at most quadratic) extension of K^σ such that $Gal(K'^{\sigma'}/K^\sigma(\delta(K'/K,\sigma)))$ is the kernel of the sign character

$$Gal(K'^{\sigma'}/K^\sigma) \longrightarrow \{\pm 1\}$$

$$s \longmapsto \begin{pmatrix} \text{sign of the permutation } t \mapsto st \\ \text{of the set } Gal(K'^{\sigma'}/K^\sigma) \end{pmatrix}$$

This and the condition that $\delta(K'/K,\sigma)^2 \in (K^\sigma)^*$ characterize the numbers defined by 1.4.5.

Furthermore, by the classical theory of the discriminant, there is an ideal $\mathfrak{b}$ of K^σ such that

$$\delta(K'/K,\sigma)^2 \cdot \mathcal{O}_{K^\sigma} = \vartheta_{K'^{\sigma'}/K^\sigma} \cdot \mathfrak{b}^2 ,$$

where $\mathcal{O}_{K^\sigma}$ is the ring of integers of K^σ and $\vartheta_{K'^{\sigma'}/K^\sigma}$ is the relative discriminant ideal of $K'^{\sigma'}$ over K^σ.

1.5 Examples

1.5.0 Let $n \in \mathbb{Z}$ and consider the n-th Tate motive $\mathbb{Q}(n)$ in $\mathcal{M}_{\mathbb{Q}}$ - cf. I,2.1 and I,2.2, Step 3. The comparison isomorphism

$$\mathbb{Q}_B(n) \otimes \mathbb{C} \xrightarrow{\sim} \mathbb{Q}_{DR}(n) \otimes \mathbb{C}$$

is simply the identity on $\mathbb{C}$. Rational bases are, say, $(2\pi i)^n$ on the left, and 1 on the right. Therefore

$$p(\mathbb{Q}(n)) = (2\pi i)^n \in \mathbb{C}^* .$$

1.5.1 Let A be an abelian variety with complex multiplication by E (necessarily a CM field) defined over K , as in I,1.1. Assume that the galois closure of E over $\mathbb{Q}$ can be embedded into K . Let $\sigma \in \mathrm{Hom}(K,\mathbb{C})$ and $\tau \in \mathrm{Hom}(E,\mathbb{C})$, and recall the Hodge exponents $n(\sigma,\tau)$ of $H_1(A)$ defined in I,1.7. Then there is a (holomorphic or antiholomorphic) 1-form

$$0 \neq \omega_{\sigma,\tau} \in H^1_{DR}(A^\sigma/K^\sigma) \cap H^{-n(\sigma,\tau),1+n(\sigma,\tau)}$$

such that, for all $e \in E \hookrightarrow \mathrm{End}_{K^\sigma}(A^\sigma)$, one has

$$e^*(\omega_{\sigma,\tau}) = e^\tau \cdot \omega_{\sigma,\tau} .$$

(Note that, by assumption on K and E , $e^\tau \in K^\sigma$.)

Choosing any nonzero rational cycle γ_σ , so that $H_1^\sigma(A)=E\cdot\gamma_\sigma$, we find that

(1.5.2) $$p(H_1(A);\sigma,\tau) = \int_{\gamma_\sigma} \omega_{\sigma,\tau} ,$$

up to the usual undeterminacy.

By the definition of an abelian variety, A admits a polarization, <u>i.e.</u>, a correspondence

$$\psi : H^1(A) \times H^1(A) \to \mathbb{Q}(-1) ,$$

and its Rosati involution necessarily induces complex conjugation c on E . So, ψ gives an E semilinear isomorphism

(1.5.3) $$H^1(A) \cong H_1(A)(-1) .$$

(We have used the fact that $H_1(A) = H^1(A)^\vee$.) Hence,

$$\begin{aligned} p(H_1(A);\sigma,\tau) &= 2\pi i \cdot p(H_1(A)(-1);\sigma,\tau) \\ &= 2\pi i \cdot p(H^1(A);\sigma,\tau c) \qquad (1.5.4) \\ &= \frac{2\pi i}{p(H_1(A);\sigma,\tau c)} \end{aligned}$$

This relation generalizes Legendre's period relation from elliptic curves to abelian varieties - here in the case of complex multiplication. It allows to express all periods of A in terms of $2\pi i$ and periods of <u>holomorphic</u> 1-forms on A .

There is also the following relation, which is valid under quite general circumstances: see 1.6.6 below.

(1.5.5) $$p(H_1(A);c\sigma,\tau) = \overline{p(H_1(A);\sigma,c\tau)} .$$

Since E is a CM field, the right hand side may also be written as the complex conjugate of $p(H_1(A);\sigma,\tau c)$.

<u>1.6 Definition of $c^\pm(M)$</u>

<u>1.6.0</u> It follows from 1.4.7, 1.5.4, and 1.5.5 that, for any abelian variety A as in 1.5.2, with real periods $p(H_1(A);\sigma,\tau)$, the period $p(R_{K/\mathbb{Q}}H_1(A))$ is essentially $2\pi i$. We shall

now recall Deligne's device to separate holomorphic from anti-holomorphic periods over $\mathbb{Q}$. We generalize it very slightly by working over a totally real field K.

1.6.1 So, let M be, as before, a motive with coefficients in E defined over K; but assume that K is a totally real number field. Then, for every $\sigma : K \hookrightarrow \mathbb{R} \subset \mathbb{C}$, the realization $H_\sigma(M)$ carries the involution

$$F_\infty : H_\sigma(M) \circlearrowleft$$

induced by complex conjugation on $H_{\mathbb{A}^f}(M)$, or directly by

$$1 \times c : M \times_{K,\sigma} \mathbb{C} \to M \times_{K,c\sigma} \mathbb{C} = M \times_{K,\sigma} \mathbb{C} .$$

Clearly, $F_\infty \otimes 1_{\mathbb{C}}(H_\sigma^{pq}) = H_\sigma^{qp}$.

Write the $+$ (resp. $-$) eigenspace of F_∞ as

$$H^{\pm}(M) = \{\gamma \in H_\sigma(M) \mid F_\infty\gamma = \pm\gamma\} .$$

Both are E submodules of $H_\sigma(M)$: being defined over K, the action of E on M commutes with F_∞. Then

$$\dim_{E\otimes\mathbb{C}}((H_\sigma^{\pm}(M) \otimes \mathbb{C}) \cap \bigoplus_{p\neq q} H_\sigma^{pq}) = \dim_{E\otimes\mathbb{C}}(\bigoplus_{p>q} H_\sigma^{pq}) .$$

In order to include the H_σ^{pp}'s we impose the

1.6.2 Assumption: There is $\pi \in \{+,-\}$ such that, for all $\sigma : K \hookrightarrow \mathbb{C}$, the involution $F_\infty \otimes 1_{\mathbb{C}}$ acts as multiplication by $\pi 1$ on all spaces H_σ^{pp} that occur in the Hodge decomposition of $H_\sigma(M)$.

This hypothesis will be made whenever we speak of the periods $c^{\pm}(M)$, to be defined presently.

1.6.3 The comparison isomorphism I_σ (1.2.3) transforms $F_\infty \otimes c$ on $H_\sigma(M) \otimes C$ into $1 \otimes c$ on $H_{DR}(M) \otimes_{K,\sigma} \mathbb{C}$. (See [DP], 1.4, for the proof; cf. I,7.3.1 above.) Thus $H_\sigma^{+}(M) \otimes \mathbb{R}$ is real with respect to the real structure

$I_\sigma^{-1} H_{DR}(M \times_{K,\sigma} \mathbb{R})$, and $H_\sigma^-(M)$ is purely imaginary.

Recall the comparison of the Hodge filtration on $H_{DR}(M)$ with the Hodge decomposition of $H_\sigma(M)$:

$$F^p H_{DR}(M) \otimes_{K,\sigma} \mathbb{C} = I_\sigma\left(\bigoplus_{p' \geq p} H_\sigma^{p',q'}\right) .$$

With π as in 1.6.2, define the $K \otimes E$ linear subspace $F^\pi H_{DR}(M)$ of $H_{DR}(M)$ by

$$F^\pi H_{DR}(M) \otimes \mathbb{C} = I\left(\bigoplus_{\sigma : K \hookrightarrow \mathbb{C}} \bigoplus_{p \geq q} H_\sigma^{pq}\right) ,$$

and put

$$F^{-\pi} H_{DR}(M) \otimes \mathbb{C} = I\left(\bigoplus_\sigma \bigoplus_{p > q} H_\sigma^{pq}\right) .$$

Note that, if $H_\sigma(M)$ is homogeneous of even weight $w = 2p$, then $F^\pi H_{DR}(M) = F^p H_{DR}(M)$ and $F^{-\pi} H_{DR}(M) = F^{p+1} H_{DR}(M)$. If w is odd, then 1.6.2 is vacuous, and $F^+ H_{DR}(M) = F^- H_{DR}(M) = F^{(w+1)/2} H_{DR}(M)$.

Then a count of dimensions shows that the isomorphism I of 1.1.1 induces isomorphisms of free $K \otimes E \otimes \mathbb{C}$ modules

(1.6.4) $$I^\pm : \bigoplus_\sigma H_\sigma^\pm(M) \otimes \mathbb{C} \xrightarrow{\sim} H_{DR}^\pm(M) \otimes \mathbb{C} ,$$

where we have put

$$H_{DR}^\pm(M) = H_{DR}(M)/F^\mp H_{DR}(M) .$$

1.6.5 In analogy to $p(M)$ above, we define (under the assumption that K is totally real, and that 1.6.2 holds):

$$c^\pm(M) \in (K \otimes E \otimes \mathbb{C})^* / (E^*)^{\mathrm{Hom}(K,\mathbb{C})} \cdot (K \otimes E)^*$$

to be the determinant of $I^\pm$, computed with respect to E bases of the $H_\sigma^\pm(M)$'s on the left, and a $K \otimes E$ basis of $H_{DR}^\pm(M)$, on the right.

As with $p(M)$, write the coordinates:

$$c^{\pm}(M) = (c^{\pm}(M;\sigma,\tau))_{\sigma,\tau} \in (\mathbb{C}^*)^{\mathrm{Hom}(K,\mathbb{C}) \times \mathrm{Hom}(E,\mathbb{C})} .$$

Note that, if $K = \mathbb{Q}$, one has simply

$$c^{\pm}(M) \in (E \otimes \mathbb{C})^*/E^* .$$

1.6.6 Remark. In the general situation of 1.1, with no assumption on K or E , we have

$$F_\infty : H_\sigma(M) \to H_{c\sigma}(M) ,$$

for all $\sigma : K \hookrightarrow \mathbb{C}$. So, one can choose E bases $\{\gamma_{i\sigma}\}$ of $H_\sigma(M)$ such that $F_\infty(\gamma_{i\sigma}) = \gamma_{i(c\sigma)}$, for all σ . Since $1 \otimes c$ coincides with F_∞ on $H_{DR}(M)$ - via I - one sees that

$$p(M;c\sigma,c\tau) = \overline{p(M;\sigma,\tau)} ,$$

up to the usual indeterminacy .

1.7 c and p

Consider the following

1.7.0 Situation. Let K and E be totally imaginary number fields, and M a motive with coefficients in E defined over K , of rank r over E . Let K_o be a totally real subfield of K (e.g., $K_o = \mathbb{Q}$), and put $M_o = R_{K/K_o}(M)$. Assume (for simplicity) that M is homogeneous of weight w . Suppose that, for all $\sigma \in \mathrm{Hom}(K,\mathbb{C})$ and $\tau \in \mathrm{Hom}(E,\mathbb{C})$, the subspace

$$H_\sigma(M) \otimes_{E,\tau} \mathbb{C} \subset H_\sigma(M) \otimes \mathbb{C}$$

is of pure Hodge type $(n(\sigma,\tau),w-n(\sigma,\tau))$, for some $n(\sigma,\tau) \in \mathbb{Z}$. (If $r = 1$, this is automatic and has been used before; for instance, in the proof of I,5.1.) Note that, for all $s \in \mathrm{Aut}\,\mathbb{C}$, one has $n(s\sigma,s\tau) = n(\sigma,\tau)$. Finally, assume that $H_\sigma^{\frac{w}{2},\frac{w}{2}} = 0$, for all $\sigma : K \hookrightarrow \mathbb{C}$.

Under these circumstances we shall now compute the periods $c^{\pm}(M_o)$ in terms of $p(M)$, using basically the same method as in 1.4.2 above. (Cf. [DP], 8.16.)

1.7.1 Let $\sigma_o \in \mathrm{Hom}(K_o,\mathbb{C})$, and start by choosing an E basis of $H^+_{\sigma_o}(M_o) = [\bigoplus_{\sigma|K_o=\sigma_o} H_\sigma(M)]^+$: Denote by $S(\sigma_o)$ the set $\{\sigma : K \hookrightarrow \mathbb{C} | \sigma|_{K_o} = \sigma_o\}$ modulo the action of complex conjugation c. For each $\bar{\sigma} = \{\sigma, c\sigma\} \in S(\sigma_o)$, choose a basis $\{\gamma_{i\sigma}\}_{i=1,\dots,r}$ of $H_\sigma(M)$ over E, and take

$$\{\gamma_{i\sigma} + F_\infty(\gamma_{i\sigma}) | i=1,\dots,r; \{\sigma,c\sigma\} \in S(\sigma_o)\}$$

as E basis of $H^+_{\sigma_o}(M_o)$. - Note that $F_\infty(\gamma_{i\sigma}) \in H_{c\sigma}(M)$ - see remark 1.6.6 -, and that our construction does not depend on the choice of the representatives $\sigma \in \bar{\sigma}$.

1.7.2 There is a unique direct factor $(K\otimes E)^+$ of $K\otimes E$ such that

$$(K\otimes E)^+ \otimes \mathbb{C} = \mathbb{C}^{\{(\sigma,\tau)|n(\sigma,\tau) < \frac{w}{2}\}} \subset K\otimes E\otimes \mathbb{C} ,$$

with $n(\sigma,\tau)$ as in 1.7.0; and the quotient $H^+_{DR}(M_o)$ of $H_{DR}(M_o)$ is isomorphic (as $K_o \otimes E$ module) to the direct factor

$$H_{DR}(M) \otimes_{K\otimes E} (K\otimes E)^+ \text{ of } H_{DR}(M) .$$

Starting from a basis $\{\omega_i\}$ of $H_{DR}(M)$ over $K\otimes E$, with components

$$\omega_i = (\omega_{i,\sigma,\tau})_{\sigma,\tau} \in \bigoplus_{\sigma,\tau} H_{DR}(M) \otimes_{K\otimes E,\sigma\otimes\tau} \mathbb{C} = H_{DR}(M)\otimes\mathbb{C} ,$$

consider the $E\otimes K_o \otimes_{K_o,\sigma_o} \mathbb{C}$ basis of $H^+_{DR}(M_o) \otimes_{K_o,\sigma_o} \mathbb{C}$

$$\{\omega_{i,\bar{\sigma}} \mid i=1,\dots,r;\ \bar{\sigma} \in S(\sigma_o)\} ,$$

where $\omega_{i,\bar{\sigma},\tau} = \omega_{i,\sigma,\tau}$, if $\sigma \in \bar{\sigma}$ and $n(\sigma,\tau) < \frac{w}{2}$.

Then we find

$$(1.7.3) \qquad c^+(M_o;\sigma_o,\tau) = \Big(\prod_{\substack{\sigma|_{K_o}=\sigma_o \\ n(\sigma,\tau)<\frac{w}{2}}} p(M;\sigma,\tau)\Big)\cdot D^+(\sigma_o,\tau) \quad ,$$

where $D^+ \in (K_o \otimes E \otimes \mathbb{C})^*$ - well determined up to $(E^*)^{\mathrm{Hom}(K_o,\mathbb{C})} \cdot (K_o \otimes E)^*$ - is the determinant of the identity on $H^+_{DR}(M_o) \otimes \mathbb{C}$, computed with respect to the basis $\{\omega_{i,\bar{\sigma}}\}_{i,\sigma}$, on the left, and some $K_o \otimes E$ basis of $H^+_{DR}(M_o)$, on the right. To compute D^+ note first that $(K\otimes E)^+$ is, in fact, a free $K_o \otimes E$ module, because K_o is totally real. Pick a basis of it:

$$\{e_j \mid j=1,\ldots,\frac{[K:K_o]}{2}\} .$$

If $\{\omega_i | i=1,\ldots,r\}$ is the $K \otimes E$ basis of $H_{DR}(M)$ used before, with ω_i projecting to ω_i^+ in $H^+_{DR}(M_o)$, then $\{e_j\, \omega_i^+\}_{j,i}$ is a $K_o \otimes E$ basis of $H^+_{DR}(M_o)$. Thus, writing $\delta^+ \in (K_o \otimes E \otimes \mathbb{C})^*$ - well determined, as usual, up to $(E^*)^{\mathrm{Hom}(K_o,\mathbb{C})} \cdot (K_o \otimes \mathbb{C})^*$ - the array with components

$$(1.7.4) \qquad \delta^+(\sigma_o,\tau) = \det\Big(e_j^\sigma \;\Big|\; \begin{matrix} j=1,\ldots,[K:K_o]/2; \\ \sigma|_{K_o}=\sigma_o,\; n(\sigma,\tau)<\frac{w}{2} \end{matrix}\Big) ,$$

we see that

$$(1.7.5) \qquad D^+ = (\delta^+)^{\varepsilon(M)} \quad ,$$

where $\varepsilon(M) = r (\mathrm{mod}\ 2)$.

<u>1.7.6</u> Like in 1.4.8, let us also give an abstract characterization of δ^+ - cf. [HS], 4.5.

Start with one fixed $\tau_o \in \mathrm{Hom}(E,C)$. For each $\sigma_o : K_o \hookrightarrow \mathbb{C}$ independently, choose $\delta^+(\sigma_o,\tau_o) \in \mathbb{C}^*$ such that the group

$$\left\{ s \in \mathrm{Gal}(\overline{\mathbb{Q}}/K^{\sigma_o}) \,\middle|\, \begin{array}{l} \text{for all } \sigma : K \hookrightarrow \mathbb{C} \text{ with } \sigma_{|K_o} = \sigma_o : \\ n(s\sigma,\tau_o) < \frac{w}{2} \Leftrightarrow n(\sigma,\tau_o) < \frac{w}{2} \end{array} \right\}$$

acts on $\delta^+(\sigma_o,\tau_o)$ via the sign character

$$s \longmapsto \left(\begin{array}{l} \text{sign of the permutation of the set} \\ \{\sigma_{|K_o} = \sigma_o \,|\, n(\sigma,\tau_o) < \frac{w}{2}\} \text{ induced by } s \end{array} \right) .$$

It remains to define $\delta^+(\sigma_o,\rho\tau_o)$, for all $\sigma_o \in \mathrm{Hom}(K_o,\mathbb{C})$ and $\rho \in \mathrm{Gal}(\overline{\mathbb{Q}}/\mathbb{Q})$. We put

$$\delta^+(\sigma_o,\rho\tau_o) = \epsilon(\rho;\sigma_o) \cdot \delta^+(\rho^{-1}\sigma_o,\tau_o)^\rho ,$$

where the signs $\epsilon(\rho;\sigma_o)$ are defined as follows. For each $\sigma_o : K_o \hookrightarrow \mathbb{C}$, choose an ordering of the set of infinite places of K lying above the place of K_o induced by σ_o . Note that, for any τ , the set $\{\sigma_{|K_o} = \sigma_o | n(\sigma,\tau) < \frac{w}{2}\}$ is in bijection with the places of K above σ_o . Then $\epsilon(\rho;\sigma_o)$ is the sign of that permutation of the places above σ_o which transforms the chosen ordering into the image under ρ of the chosen ordering on the set of places above $\rho^{-1}\sigma_o$. - The choices made are compensated by the indeterminacy of δ^+ up to $(E^*)^{\mathrm{Hom}(K_o,\mathbb{C})}$.

1.7.7 Let us now derive the analogue of 1.7.3 for c^- . In the notation of 1.7.1,

$$\{\gamma_{i\sigma} - F_\infty(\gamma_{i\sigma}) | i=1,\ldots,r; \{\sigma,c\sigma\} \in S(\sigma_o)\}$$

is a basis of $H^-_{\sigma_o}(M_o)$ over E . Note that here we have to take a particular choice of representatives $\{\sigma\}$ of $S(\sigma_o)$. One way to do this - which we adopt - is again to fix one embedding $\tau_o : E \hookrightarrow \mathbb{C}$, and to use the set

$$\{\sigma | \sigma_{|K_o} = \sigma_o \text{ and } n(\sigma,\tau_o) < \tfrac{w}{2}\} .$$

Define accordingly,

$$\widetilde{\omega}_{i,\overline{\sigma}} \in H^-_{DR}(M_o) \otimes_{K_o,\sigma_o} \mathbb{C} = H^+_{DR}(M_o) \otimes_{K_o,\sigma_o} \mathbb{C}$$

by their components:

$$\tilde{\omega}_{i,\bar{\sigma},\tau} = \begin{cases} \omega_{i,\sigma,\tau}, & \text{if } \sigma \in \bar{\sigma} \text{ and } n(\sigma,\tau) < \frac{w}{2} > n(\sigma,\tau_o) \\ -\omega_{i,\sigma,\tau}, & \text{if } \sigma \in \bar{\sigma} \text{ and } n(\sigma,\tau) < \frac{w}{2} < n(\sigma,\tau_o) \end{cases} .$$

This gives

$$c^-(M_o;\sigma_o,\tau) = \Big(\prod_{\substack{\sigma|K_o = \sigma_o \\ n(\sigma,\tau) < \frac{w}{2}}} p(M;\sigma,\tau)\Big)\cdot D^-(\sigma_o,\tau) , \tag{1.7.8}$$

where D^- is defined like D^+ in 1.7.3, with $\{\omega_{i,\bar{\sigma}}\}$ replaced by $\{\tilde{\omega}_{i,\bar{\sigma}}\}$. In particular, like in 1.7.5,

$$D^- = (\delta^-)^{\varepsilon(M)} , \tag{1.7.9}$$

where the quotient δ^+/δ^- is given - up to the usual indeterminacy - by the rule

$$\frac{\delta^+(\sigma_o,\tau)}{\delta^-(\sigma_o,\tau)} = (-1)^{\#\{\sigma|K_o = \sigma_o | n(\sigma,\tau) < \frac{w}{2} < n(\sigma,\tau_o)\}} , \tag{1.7.10}$$

for all $\sigma_o \in \mathrm{Hom}(K_o,\mathbb{C})$, $\tau \in \mathrm{Hom}(E,\mathbb{C})$, and τ_o as fixed above.

<u>1.7.11</u> <u>Corrigendum.</u> Formula 1.7.10 emends our foolish negligence at the end of the proof of [GS], 9.3, and again in [GS'], 3.3. There we asserted, for $K_o = \mathbb{Q}$, K quadratic, and $r=1$, that $c^+ = c^-$. In "proving" Deligne's conjecture in that case, we compensated this mistake by overlooking the fact that the complex conjugate of $2\pi i$ is $-2\pi i$, in the application of [DP], 5.18. The same false replacement of c^- for c^+ still slipped into [HS], formula 11 - cf. instead, 3.1. below -, where it was finally caught by Blasius.

<u>1.7.12</u> <u>Lemma</u>

(i) <u>$\delta^{\pm}$ depend only on</u> K_o, K, E, <u>and the family of "CM-types" of</u> K $(\{\sigma \in \mathrm{Hom}(K,\mathbb{C}) | n(\sigma,\tau) < \frac{w}{2}\})_{\tau : E \hookrightarrow \mathbb{C}}$.

(ii) $(\delta^{\pm})^2 \in (E^*)^{\mathrm{Hom}(K_o,C)} (K_o \otimes E)^*$.

(iii) Let K'/K be an extension of degree n , and denote by $\delta'^{+} \in (K_o \otimes E \otimes \mathbb{C})^*$ the δ^{+}-factor relative to K'/K_o , and the exponents $n(\sigma',\tau) = n(\sigma|_{K},\tau)$. Then, for all $\sigma_o \in \mathrm{Hom}(K_o,\mathbb{C})$ and $\tau \in \mathrm{Hom}(E,\mathbb{C})$:

$$\frac{\delta'^{\pm}(\sigma_o,\tau)}{\delta^{\pm}(\sigma_o,\tau)^n} = \prod_{\substack{\sigma|_{K_o} = \sigma \\ n(\sigma,\tau)<\frac{w}{2}}} \delta(K'/K,\sigma) \, ,$$

where $\delta(K'/K)$ has been defined in 1.4.5/6.

(iv) If $K_o = \mathbb{Q}$, and K a CM field with maximal real subfield F , then - up to a factor in $(E \otimes 1)^*$ - ,

$$\delta^{+} = \sqrt{\mathrm{disc}(F)} \, ,$$

where we take one and the same root of the absolute discriminant of F , for all embeddings of E into $\mathbb{C}$.

Parts (i) and (ii) are plain. In part (iii) note that the various indeterminacies actually do work out: If, for $k \in K^*$, $\delta(K'/K,\sigma)$ is replaced by $k^{\sigma}\,\delta(K'/K,\sigma)$, then we obtain in the formula the factor

$$\prod_{\substack{\sigma|_{K_o} = \sigma_o \\ n(\sigma,\tau)<\frac{w}{2}}} k^{\sigma} \in (K_o^{\sigma_o}\cdot E^{\tau})^* \subset \mathbb{C}^* \, .$$

The proof of (iii) is straightforward. (iv) is also easy to prove once one observes that $\mathrm{Hom}(F,\mathbb{C})$ naturally identifies itself with $S(\mathrm{id}_{\mathbb{Q}})$. - Cf. [DP], 8.17; and 1.4.8 above.

1.8 Application to Hecke characters

We now resume the discussion of the "unique" motive $M(\chi)$ in $\mathcal{M}_K^{av}(E)$ which we have attached in chapter I (I,4, I,5; also

I, 6.5.6) to a given algebraic Hecke character χ of K with values in E . Write its periods as

$$p(\chi) = p(M(\chi)) \in (K \otimes E \otimes \mathbb{C})^*/(E^*)^{\mathrm{Hom}(K,\mathbb{C})} (K \otimes E)^* ,$$

with components $p(\chi;\sigma,\tau)$. And <u>if</u> $R_{K/\mathbb{Q}}M(\chi)$ <u>satisfies hypothesis 1.6.2,</u> it makes sense to write

$$c^{\pm}(\chi) = c^{\pm}(R_{K/\mathbb{Q}}M(\chi)) \in (E \otimes \mathbb{C})^*/E^* ,$$

with components $c^{\pm}(\chi;\tau)$.

Recall that, by definition (I, 3.3), $M(\chi)$ is of rank 1 over E , and that its Hodge decomposition is given by the invariants $n(\sigma,\tau)$ attached to χ in 0,4, the weight of the Hodge structure being the weight w of χ : see I, 6.1.5. It follows that, if none of the $n(\sigma,\tau)$'s equals $\frac{w}{2}$, then

$$(1.8.1) \qquad c^{\pm}(\chi) = \delta^{\pm}(\chi) \Big(\prod_{n(\sigma,\tau)<\frac{w}{2}} p(\chi;\sigma,\tau) \Big)_{\tau} .$$

In fact, this is just a reformulation of 1.7.3, resp. 1.7.8, with $\delta^{\pm}(\chi) \in (E \otimes \mathbb{C})^*/E^*$ equal to the factor given by 1.7.4, resp. 1.7.10, relative to the data $\mathbb{Q}$, K, E, and the $n(\sigma,\tau)$'s of χ - see 1.7.12 (i).

<u>1.8.2</u> In a nutshell, the observation which is basic to our work is that, by theorem I, 5.1, <u>all these periods do not depend on the particular geometric construction of a motive</u> $M(\chi)$ <u>in</u> $\mathcal{M}_K^{av}(E)$ <u>attached to</u> χ .

As a first illustration of this principle we shall now give a list of six basic properties of the periods $p(\chi)$ all of which follow from two different ways of writing the $M(\chi)$ in question. These isomorphisms of motives are all easily checked on the λ-adic representations, i.e. precisely, by verifying that the motives on both sides of what we shall write as an equality are motives for one and the same character. In each case it is indicated, how the period relation follows from the corresponding isomorphism of motives. - The reader will notice the analogy of our list to the table in [DB], § 2.

<u>1.8.3</u> Let χ and χ' be two algebraic Hecke characters of K with values in E. Then

$$M(\chi\cdot\chi') = M(\chi) \otimes_E M(\chi') \; ; \quad p(\chi\cdot\chi') = p(\chi)\cdot p(\chi') \; .$$

Since the motives are of rank 1 over E, it is clear that the isomorphism on the left implies the simple period relation on the right.

<u>1.8.4</u> Let E be $\mathbb{Q}$, and denote by $\mathbb{N}$ the absolute norm of ideals of K. Then, for all $n \in \mathbb{Z}$,

$$M(\mathbb{N}^n) = \mathbb{Q}(-n) \times_{\mathbb{Q}} K \; ; \quad p(\mathbb{N}^n) = (2\pi i)^{-n} \; .$$

This follows from 1.5.0 and 1.4.1. Note that, if K is also $\mathbb{Q}$, then $(2\pi i)^{-n} = c^{(-)^n}(\mathbb{N}^n)$ - see I, 2.1 for the action of F_∞ on the Tate motive.

<u>1.8.5 Remark.</u> If χ is the Hecke character of $H_1(A)$, for A an abelian variety with complex multiplication, like in I,1, then $\chi\cdot\bar{\chi} = \mathbb{N}^{-1}$. Thus 1.8.3 and 1.8.4 reprove "Legendre's period relation", 1.5.4.

<u>1.8.6</u> Let E'/E be a finite extension, χ an algebraic Hecke character of K with values in E. Then, with i the inclusion $E^* \hookrightarrow E'^*$ (also viewed as homomorphism of algebraic groups),

$$M(i\circ\chi) = M(\chi) \otimes_E E' \; ; \quad p(i\circ\chi) = i(p(\chi)) \; .$$

This is just an application of 1.3.1.

<u>1.8.7</u> Let again E'/E be a finite extension, but let a Hecke character χ' of K with values in E' be given. Then, denoting $N_{E'/E}$ the norm homomorphism $E'^* \to E^*$,

$$M(N_{E'/E}\circ\chi') = \det_E(M(\chi)|_E) \; ; \quad p(N_{E'/E}\circ\chi) = N_{E'/E}(p(\chi)) \; .$$

Recalling that $\det_E$ was defined in 1.1.2 (and the restriction of coefficients $|_E$ in I, 3.0), the period relation is implied by the isomorphism, because of 1.1.3 and 1.3.2.

1.8.8 Let K'/K be a finite extension, and denote by $N_{K'/K}$ the relative norm, on ideals of K'. Then, for χ an algebraic Hecke character of K, and $j : K^* \hookrightarrow K'^*$ the inclusion,

$$M(\chi \circ N_{K'/K}) = M(\chi) \times_K K' \; ; \quad p(\chi \circ N_{K'/K}) = j(p(\chi)) \, ,$$

as follows from 1.4.1.

1.8.9 Given again a finite extension K'/K, but a Hecke character χ' of K' with values in E, write now j for the inclusion of ideals of K into ideals of K', and $N_{K'/K}$ for the norm $K'^* \to K^*$. Let $\epsilon_{K'/K}$ be the finite order character of K which, via Artin reciprocity, corresponds to the character

$$\epsilon_{K'/K} : \mathrm{Gal}(\overline{K}/K) \to \{\pm 1\}$$
$$s \longmapsto \begin{pmatrix} \text{sign of the permutation of the set} \\ G(\overline{K}/K)/G(\overline{K}/K') \text{ given by } s \end{pmatrix}$$

Then the following isomorphism of motives is an easy generalization of Prop. 3.2 of [Mar], p. 35 f.

$$M(\epsilon_{K'/K} \cdot (\chi' \circ j)) = \det_E(R_{K'/K} M(\chi')) ; \; p(\chi' \circ j) = N_{K'/K}(p(\chi')) \, .$$

The period relation follows from 1.1.3, 1.4.2, 1.8.3, and the fact that $p(\epsilon_{K'/K}) = \delta(K'/K)$ which will be proved in 3.2 below.

1.8.10 It is usually clear how these formulas for the periods p can be used to derive relations between periods $c^{\pm}$, using 1.7, resp. 1.8.1. However, care must be taken not to apply 1.8.3 inside 1.8.1, unless both factors in question satisfy the conditions of 1.7.0 with the same system of "CM-types" $\{\sigma \,|\, n(\sigma,\tau) < \frac{w}{2}\}_{\tau \in \mathrm{Hom}(E,\mathbb{C})}$.

See section 3 for the most common illustrations of this problem.

A relation deduced, with no such difficulty, by combining 1.7.12 (i), 1.7.12 (iii), 1.8.3 and 1.8.8, is the formula which plays a crucial role in [HS]: If K, K' and χ are like in 1.8.8, and $n = [K':K]$, then

$$(1.8.11) \qquad \frac{c^+(\chi \circ N_{K'/K})}{c^+(\chi^n)} = \Big(\prod_{n(\sigma,\tau) < \frac{w}{2}} \delta(K'/K,\sigma) \Big)_\tau \cdot \delta^+(\chi)^{n-1} \ .$$

2. Periods and L-values

As usual, let K and E be number fields, and χ a Hecke character of K with values in E.

2.0 Let $\tau \in \mathrm{Hom}(E,\mathbb{C})$. An integer s is called critical for (the L-function of) χ^τ, if the Γ-factors on both sides of the functional equation of $L(\chi^\tau,.)$ do not have a pole at s. It is an easy exercise to work out what this means, using the formulas in 0, § 6 - cf. [DP], 1.3, 3, 8.15 - :

2.0.1 If $\{\sigma, c\sigma\}$, for $\sigma \in \mathrm{Hom}(K,\mathbb{C})$, induces a complex place of K, then $n(\sigma,\tau)$ has to be different from $\frac{w}{2}$, for a critical s to exist. Thus critical integers (for Hecke characters) can only occur, if K is totally real or totally imaginary, and in the latter case, one has a disjoint union

$$\mathrm{Hom}(K,\mathbb{C}) \times \mathrm{Hom}(E,\mathbb{C}) = \{(\sigma,\tau) \mid n(\sigma,\tau) < \tfrac{w}{2}\} \,\dot{\cup}\, \{(\sigma,\tau) \mid n(\sigma,\tau) > \tfrac{w}{2}\} \ .$$

2.0.2 If K is totally real, the character $\mu \mathbb{N}^n$ - with μ of finite order and $n \in \mathbb{Z}$ - admits critical integers s if and only if, for all infinite places v of K, the constants ε_v defined for μ_v^τ in 0, § 6 are equal to, say, $\varepsilon \in \{0,1\}$. Then the set of all critical s for $\mu \mathbb{N}^n$ is

$$\{s > n \mid s \equiv n+\varepsilon (\mathrm{mod}\ 2)\} \cup \{s \leq n \mid s \not\equiv n+\varepsilon (\mathrm{mod}\ 2)\} \ .$$

2.0.3 If K is totally imaginary, critical s exist for χ^τ, if and only if, for all σ,

$$n(\sigma,\tau) < \tfrac{w}{2} \qquad \text{or} \qquad n(\sigma,\tau) > \tfrac{w}{2} \ .$$

If this is so, then the set of critical integers for χ - independently of τ - is the interval

$$\Big(\sup_{n(\sigma,\tau)<\frac{w}{2}} n(\sigma,\tau), \quad \inf_{n(\sigma,\tau)>\frac{w}{2}} n(\sigma,\tau) \Big].$$

In all cases, we can therefore say "critical for χ", independently of τ. - Recall the notation $L^*(\chi,s)$ from 0, §6; and $c^+(\chi)$ from 1.8.

<u>2.1 Theorem</u> [Siegel, Blasius, Harder]. <u>If</u> s <u>is critical for</u> χ, <u>then</u>

$$\frac{L^*(\chi,s)}{c^+(\chi\cdot N^{-s})} \in E \otimes 1 \hookrightarrow E \otimes \mathbb{C}.$$

In other words, Deligne's conjecture [DP], 2.8 is true "for all algebraic Hecke characters".

We have discussed the history as well as the important steps of the proof of this deep theorem in [HS], see especially §5. Let us just recall here that Siegel's contribution concerns the case when K is totally real; Blasius [Bl] proves it for K a CM field, refining Shimura's groundbreaking proof of this case up to factors in $\overline{\mathbb{Q}}^*$ - see [ShiA], [DP], 8.21. Finally, Harder's methods produce relations between L-values which allow one to carry over Blasius' result from CM fields to arbitrary totally imaginary fields by virtue of the analogous relations for the c^+-periods (cf. 3.2.7 below). The considerable technicalities of his argument have not been written up yet in the general case. Cf. However [Ha].

<u>2.2 Remark.</u> One of the key constructions in [Bl] is the construction, for $M = R_{K/\mathbb{Q}}M(\chi)$, of a motive that plays a role analogous to $\det_E M$ in 1.1.3, with p replaced by c^+. One can use this construction to derive all the formulae relative to $c^{\pm}$ which we have presented.

3. Twisting

3.0 We continue to consider an algebraic Hecke character χ of the number field K with values in the number field E. Assume that K is totally imaginary, and that χ admits some critical integer s - see 2.0.1.

3.1 Write $\mathbb{N}$ for the absolute norm (of ideals of K). Then

$$(3.1.1) \qquad c^{\pm}(\chi\mathbb{N}^{-1}) = (2\pi i)^{\frac{[K:\mathbb{Q}]}{2}} c^{\mp}(\chi).$$

[This rectifies formula (11) of [HS] - cf. 1.7.11.]

3.1.1 is a special case of [DP], 5.1.8, which follows from the characterizing properties of $\mathbb{Q}(1)$ - see I, 2.1. In trying to derive 3.1.1 from 1.8.1 (for $\chi\mathbb{N}^{-1}$), 1.8.3, and 1.8.4, the subtlety is that $c^{\pm}$ becomes $c^{\mp}$ because, for $\gamma_\sigma \in H_\sigma(M(\chi))$ and $\alpha \in \mathbb{Q}_B(1) \otimes E$,

$$\gamma_\sigma \otimes_E \alpha + F_\infty(\gamma_\sigma \otimes_E \alpha) = (\gamma_\sigma - F_\infty(\gamma_\sigma)) \otimes_E \alpha;$$

cf. 1.7.7.

3.2 Next, let μ be a Hecke character of K of finite order, with values in E. (In view of 1.3.1, there is no loss of generality, for our period calculations, in assuming that μ and χ both take values in the same field E. This will be assumed in those of the following formulae which involve both μ and χ.) Then 1.8.3 gives:

$$(3.2.0) \qquad p(\mu\chi) = p(\mu)p(\chi).$$

We shall compute $p(\mu)$ using the explicit description of the Artin motive $M(\mu)$ given in I, 2.4.1.

3.2.1 Let F be the finite abelian extension of K corresponding to μ by class field theory. Thus, reading μ on $\Gamma = \mathrm{Gal}(\overline{K}/K)$ - for some fixed algebraic closure $\overline{K}$ of K -

via geometric Frobenii, F is the fixed field of $\ker(\mu)$. There are two actions of Γ on $F \otimes E$ inducing the natural action of Γ on $F \subset \overline{K}$:

- the trivial action of Γ on E : $(f \otimes e)^{\gamma} = f^{\gamma} \otimes e$;
- the action $(f \otimes e)^{[\gamma]} = f^{\gamma} \otimes \mu(\gamma)e$.

Clearly $y^{[\gamma]} = (1 \otimes \mu(\gamma)) \cdot y^{\gamma}$, for all $y \in F \otimes E$ and $\gamma \in \Gamma$. - We shall usually work with the first action. In particular, this is the action for which we have the

3.2.2 Lemma $\quad H^1(\Gamma/\ker(\mu),(F \otimes E)^*) = 0$.

Proof [suggested by M. Lorenz]: Let $\Delta = \Gamma/\ker(\mu)$, and write $E = \mathbb{Q}[X]/(P)$, with $P \in \mathbb{Q}[X]$ irreducible. Then $F \otimes E = F[X]/(P)$. Factor $P = P_1 \ldots P_s$ in $F[X]$. Acting on $F[X]$ through the coefficients Δ permutes the ideals $(P_1),\ldots,(P_s)$, since it stabilizes (P) . So, writing the orbits one by one, we have

$$(P) = \prod_{i=1}^{t} \prod_{j=1}^{s_i} (P_{ij}) ,$$

and for each i , Δ permutes the (P_{ij}) transitively.

Then

$$(F \otimes E)^* = \prod_{i=1}^{t} A_i^* ,$$

as Δ-module, where

$$A_i = \prod_{j=1}^{s_i} F[X]/(P_{ij}) .$$

Since $H^1(\Delta,\prod_i A_i^*) = \bigoplus_i H^1(\Delta,A_i^*)$ we are reduced to the case of a transitive Δ-action. In other words, if Δ_{i1} is the subgroup of Δ stabilizing $F[X]/(P_{i1})$, then

$$A_i^* = \mathrm{Ind}_{\Delta_{i1}}^{\Delta}((F[X]/(P_{i1}))^*) .$$

By Shapiro's lemma and Hilbert 90 ,

$$H^1(\Delta, A_i^*) = H^1(\Delta_{i1}, (F[X]/P_{i1}))^*) = 0 .$$

q.e.d.

Applying the lemma to $[\mu] \in H^1(\Gamma/\ker(\mu), (F \otimes E)^*)$, we find a unit $\xi \in (F \otimes E)^*$ such that, for all $\gamma \in \Gamma$,

(3.2.3) $$\xi^\gamma = (1 \otimes \mu(\gamma)) \ \xi \ .$$

(In fact, the lemma implies that any $0 \neq \xi \in F \otimes E$ satisfying 3.2.3 lies in $(F \otimes E)^*$:) ξ is well-determined up to a factor in $(K \otimes E)^*$. - In terms of our second action, we have

$$(\overline{K} \otimes E)^{[\Gamma]} = \xi^{-1} \ (K \otimes E) \subset \overline{K} \otimes E .$$

Now, the motive $M(\mu)$ really "is" E viewed as the one dimensional E linear representation of Γ given by μ . On the other hand, $H_{DR}(M(\mu)) = (\overline{K} \otimes E)^{[\Gamma]}$ - see I, 2.4.1 -, so ξ is a $K \otimes E$ basis of $H_{DR}(M(\mu))$. Therefore, for each $\sigma : K \hookrightarrow \overline{\mathbb{Q}} \subset \mathbb{C}$, the period $p(\mu;\sigma)$ can be computed like this: take 1 as E-basis of $H_\sigma(M(\mu)) = E$; for any extension $\widetilde{\sigma} : \overline{K} \xrightarrow{\sim} \overline{\mathbb{Q}}$ of σ , the inverse of $\xi^{\widetilde{\sigma}} = \xi^{\widetilde{\sigma} \otimes id_E} \in (\overline{\mathbb{Q}} \otimes E)^*$ is a $K^\sigma \otimes E$ basis of $H_{DR}(M(\mu)) \otimes_{K,\sigma} K^\sigma \subset H_{DR}(M(\mu)) \otimes_{K,\sigma} \mathbb{C} = E \otimes \mathbb{C}$; we find

(3.2.4) $$p(\mu;\sigma) = \xi^{\widetilde{\sigma}} .$$

Let us analyze the indeterminacy: ξ was well determined up to $(K \otimes E)^*$; on the other hand, if we pick $s\widetilde{\sigma}$ instead of $\widetilde{\sigma}$, with $s \in \mathrm{Gal}(\overline{\mathbb{Q}}/K^\sigma)$, we find,

$$\xi^{s\widetilde{\sigma}} = \xi^{\widetilde{\sigma}\widetilde{\sigma}^{-1}s\widetilde{\sigma}} = [(1 \otimes \mu(\widetilde{\sigma}^{-1}s\widetilde{\sigma})) \cdot \xi]^{\widetilde{\sigma}}$$

$$= (1 \otimes \mu(\widetilde{\sigma}^{-1}s\widetilde{\sigma})) \cdot \xi^{\widetilde{\sigma}} .$$

Thus, the array $(\xi^{\widetilde{\sigma}})_{\sigma : K \hookrightarrow \mathbb{C}} \in (K \otimes E \otimes \mathbb{C})^*$ is well determined up to a factor in $(E^*)^{\mathrm{Hom}(K,\mathbb{C})} \ (K \otimes E)^*$, and we have

(3.2.5) $$p(\mu) = (\xi^{\widetilde{\sigma}})_{\sigma \in \mathrm{Hom}(K,\mathbb{C})} .$$

This establishes in particular the formula left unproven in 1.8.9 above. - Let us restate 1.8.9 for finite order characters, using the well known role of the transfer map in class field theory:

3.2.6 Let K'/K be a finite extension, and μ' a character of finite order of K' (always with values in E). Denote by $\mathrm{Ver}^K_{K'} : \mathrm{Gal}(\overline{K}/K) \to \mathrm{Gal}(\overline{K}/K')^{ab}$ the transfer map. Then

$$p(\mu' \circ \mathrm{Ver}^K_{K'}) = N_{K'/K}(p(\mu')) .$$

This formula implies the following invariance lemma a special case of which was needed as formula (12) in [HS] - cf. also 1.8.11.

3.2.7 Lemma *Let K and χ be as in 3.0. Let K'/K be a finite extension and χ' a Hecke character of K' with values in E (like χ), such that, for all $\tau : E \hookrightarrow \mathbb{C}$,*

$$\{\sigma'|_K \mid \sigma' \in \mathrm{Hom}(K',\mathbb{C}); n'(\sigma',\tau) < \tfrac{w'}{2}\} = \{\sigma \in \mathrm{Hom}(K,\mathbb{C}) \mid n(\sigma,\tau) < \tfrac{w}{2}\} .$$

Then, for any μ' as in 3.2.6, one has

$$\frac{c^{\pm}(\mu' \cdot \chi')}{c^{\pm}((\mu' \circ \mathrm{Ver}^K_{K'})\chi)} = \frac{c^{\pm}(\chi')}{c^{\pm}(\chi)} .$$

This follows from 1.8.1, 1.7.12(i), 1.8.3, and 3.2.6. Note that, unlike 3.1, the use of 1.8.3 inside 1.8.1 is licit here because K and K' are totally imaginary. In fact, even if we had, say, $\mu' = \mu_o \circ N_{K'/K_o}$, for some totally real field K_o and with F_∞ acting on $H_{\sigma_o}(M(\mu_o))$ as -1 (i.e., μ_o involves a nontrivial sign character), no such signs would be visible over K', and $F_\infty : H_{\sigma'}(M(\mu')) \to H_{c\sigma'}(M(\mu'))$ simply identifies these two spaces.

3.2.8 We shall now develop an analogue of 3.2.6, with $N_{K'/K}$ replaced by Tate's "half transfer" - cf. I, 6.4.0. Let χ

and K be as before - but assume that K is a CM field. (We can always reduce to this case by 3.2.6: see $\mathbb{0}$, § 3.) Let $K_o \subset K$ be a totally real subfield of K. (The important case will be $K_o = \mathbb{Q}$.) Fix an embedding $\sigma_o : K_o \hookrightarrow \mathbb{C}$, and consider K as embedded into $\overline{\mathbb{Q}} \subset \mathbb{C}$, by using some fixed extension of σ_o (which will not show up in the notation). Choose a system of representatives,

$$v : \mathrm{Hom}_{K_o,\sigma_o}(K,\overline{\mathbb{Q}}) \rightarrow \mathrm{Gal}(\overline{\mathbb{Q}}/K_o^{\sigma_o})$$

in such a way that $v(c\sigma) = c\,v(\sigma)$, for c = complex conjugation. For each $\tau \in \mathrm{Hom}(E,\mathbb{C})$, define the "half transfer" map attached to χ and τ, relative to K_o, σ_o,

$$V(\cdot\,,\tau) : \mathrm{Gal}(\overline{\mathbb{Q}}/K_o^{\sigma_o}) \rightarrow \mathrm{Gal}(K^{ab}/K)$$

by the rule

$$V(s,\tau) = \prod_{\sigma|K_o = \sigma_o} [v(s\sigma)^{-1} s\, v(\sigma)]^{-n(\sigma,\tau)} (\mathrm{mod}\ \mathrm{Gal}(\overline{\mathbb{Q}}/K^{ab}))\ .$$

V is independent of the choice of v, and for all $s,t \in \mathrm{Gal}(\overline{\mathbb{Q}}/K_o^{\sigma_o})$ and $\tau \in \mathrm{Hom}(E,\mathbb{C})$, one has the cocycle relation

$$(3.2.9) \qquad V(st,\tau) = V(s,t\tau)\, V(t,\tau)\ .$$

Now, let μ be any finite order character on K with values in E. Define

$$\mu_o : \mathrm{Gal}(\overline{\mathbb{Q}}/K_o^{\sigma_o}) \rightarrow (E^*)^{\mathrm{Hom}(E,\overline{\mathbb{Q}})}$$

by the rule

$$(3.2.10) \qquad \mu_o(s) = (\mu(V(s^{-1},\tau)))_\tau\ .$$

Let $K \subset F \subset \overline{\mathbb{Q}}$ be as in 3.2.1, and define a left action of $\mathrm{Gal}(\overline{\mathbb{Q}}/K_o^{\sigma_o})$ on $\mathrm{Maps}(\mathrm{Hom}(E,\overline{\mathbb{Q}}),(F \otimes E)^*)$ by using the trivial action on E^* and the natural actions on F^* and $\mathrm{Hom}(E,\overline{\mathbb{Q}})$ - cf. 3.2.1. Then there exists a unit

$$\eta \in (F \otimes E)^{*\mathrm{Hom}(E,\overline{\mathbb{Q}})} \hookrightarrow (E \otimes \mathbb{C})^{*\mathrm{Hom}(E,\mathbb{C})}$$

such that, for all $s \in Gal(\overline{\mathbb{Q}}/K_o^{\sigma_o})$,

(3.2.11) $\quad \eta^s = (1 \otimes \mu_o(s))\cdot\eta$.

In fact, we can put (see 3.2.3/4, with $\overline{K} = \overline{\mathbb{Q}}$ and $\tilde{\sigma} = v(\sigma)$) :

(3.2.12) $\quad \eta = \Big(\prod_{\sigma|K_o = \sigma_o} p(\mu;\sigma)^{n(\sigma,\tau)}\Big)_\tau$.

η is determined by 3.2.11 up to a factor in

$$(K_o^{\sigma_o} \otimes E)^{*(G(\overline{\mathbb{Q}}/K_o^{\sigma_o})\backslash Hom(E,\overline{\mathbb{Q}}))} .$$

It is convenient to write η as a matrix $(\eta_{\tau,\tau'})$, with indices $\tau,\tau' \in Hom(E,\overline{\mathbb{Q}})$, and entries

(3.2.13) $\quad \eta_{\tau,\tau'} = \prod_{\sigma|K_o = \sigma_o} (\xi^{v(\sigma)\otimes\tau'})^{n(\sigma,\tau)} \in \overline{\mathbb{Q}}^*$.

Then, for $s \in Gal(\overline{\mathbb{Q}}/K_o^{\sigma_o})$,

(3.2.14) $\quad (\eta_{\tau,\tau'})^s = \dfrac{1}{\mu(V(s,\tau))^{s\tau'}} \; \eta_{s\tau,s\tau'}$,

and, for $\eta_\tau = (\eta_{\tau,\tau'})_{\tau'} \in (\overline{\mathbb{Q}} \otimes E)^*$, with s acting via the first action introduced in 3.2.1,

(3.2.15) $\quad (\eta_\tau)^s = (1 \otimes \mu(V(s,\tau)))^{-1}\, \eta_{s\tau}$.

What makes these formulas interesting is their connection with the periods $c^\pm$, and thereby, via 2.1, with L-values:

<u>3.3.0</u> <u>Example.</u> Let A be an abelian variety with complex multiplication by E defined over K - cf. I § 1. Call χ its Hecke character: $M(\chi) = H_1(A)$. Then by 1.8.1, 1.8.3 and 3.2.13 (with $K_o = \mathbb{Q}$), one finds for any finite order character μ of K (with values in E):

(3.3.1) $\quad \dfrac{c^\pm(\mu\cdot\chi)}{c^\pm(\chi)} = (\eta_{\tau,\tau})_\tau \in (E \otimes \overline{\mathbb{Q}})^*$.

(The justification for applying 1.8.3 inside 1.8.1 is the same as in 3.2.7: K is totally imaginary.) Thus, by 3.2.14 with s fixing τ , the τ-component $\eta_{\tau,\tau}$ of this quotient of periods generates the abelian extension of E^τ corresponding to the character $\mu^\tau(V(\cdot,\tau))$ of $\mathrm{Gal}(\bar{\mathbb{Q}}/E^\tau)$. And if, by chance, both $L^*(\mu\chi,0)$ and $L^*(\chi,0)$ are in $(E\otimes\mathbb{C})^*$, then their quotient in $(E\otimes\mathbb{C})^*$ has the same property. Thus, in particular, for all $\tau : E \hookrightarrow \bar{\mathbb{Q}}$,

$$(3.3.2)\qquad \frac{L(\mu^\tau\chi^\tau,0)}{L(\chi^\tau,0)} \in (E^\tau)^{ab} , \quad \text{and} \quad \left\{\frac{L^*(\mu\chi,0)}{L^*(\chi,0)}\right\}^{\mathrm{order}(\mu)} \in E^* \subset (E\otimes\mathbb{C})^* .$$

3.3.3 It is easy to generalize the statements of this example to arbitrary Hecke characters χ of K . Assume for simplicity, that $s = 0$ is critical for χ . Define $\tilde{V}$ to be the transfer defined by the system of invariants

$$\tilde{n}(\sigma,\tau) = \begin{cases} -1 & \text{if } n(\sigma,\tau) < 0 \\ 0 & \text{if } n(\sigma,\tau) \geq 0 \end{cases} .$$

Then 3.3.1 holds for χ , with $\eta_{\tau,\tau}$ replaced by $\tilde{\eta}_{\tau,\tau}$ - defined relative to $\tilde{\mu}_0(s) = (\mu(\tilde{V}(s^{-1},\tau)))_\tau$ instead of μ_0 . So here, too, 3.3.2 follows.

3.4 Finally, let us lift our convention 3.0, and consider the case that K is totally real (embedded into $\bar{\mathbb{Q}}$) . Assume for simplicity that $s = 0$ is critical for the character $\chi = \mu\cdot\mathbb{N}^n$. Then (2.0.2) F_∞ acts on $H_B(R_{K/\mathbb{Q}}M(\mu))$ as $(-1)^n$ if $n > 0$, and as $-(-1)^n$, if $n \leq 0$. Thus, if $n \leq 0$, putting $\pi 1 = (-1)^\epsilon$, we obtain

$$(3.4.1)\qquad \begin{cases} c^+(\mu\,\mathbb{N}^n) = 1 = c^\pi(\mathbb{N}^n) \\ c^-(\mu\,\mathbb{N}^n) = p(R_{K/\mathbb{Q}}M(\chi)) = p(\epsilon_{K/\mathbb{Q}}\cdot(\mu\circ \mathrm{Ver}_K^{\mathbb{Q}})\cdot\mathbb{N}^n) . \end{cases}$$

In the case $n > 0$ the signs get reversed, and we find, for $m = -n > 0$,

$$(3.4.2)\qquad \frac{L(\mu,m)}{(2\pi i)^m} = \frac{c^+(\mu\,\mathbb{N}^n)}{c^\pi(\mathbb{N}^n)} = p(\epsilon_{K/\mathbb{Q}}\cdot(\mu\circ \mathrm{Ver}_K^{\mathbb{Q}})) .$$

Since the construction of ξ , such that 3.2.3/4 hold, clearly works over all base fields, we get in particular that $Gal(\overline{\mathbb{Q}}/\mathbb{Q})$ acts on $p(\epsilon_{K/\mathbb{Q}}\cdot(\mu\circ Ver_K^{\mathbb{Q}})) \in (E \otimes \overline{\mathbb{Q}})^*$ via the character $\epsilon_{K/\mathbb{Q}}\cdot(\mu\circ Ver_K^{\mathbb{Q}})$.

But for all Dirichlet characters of $\mathbb{Q}$, such elements are classically given by Gauss sums, and more precisely by their "root numbers"; see [DP], 6.4, 6.5. This most incredible coincidence does NOT repeat itself over algebraic number fields K different from $\mathbb{Q}$! In fact, the components of $p(\mu)$ generate the corresponding abelian extensions of K , and can therefore not all lie in $\mathbb{Q}^{ab}$. - The last sentence of [HS], § 4 is therefore INCORRECT - and should never have been put in there in the first place.

3.5 Let K and E be arbitrary number fields.

3.5.1 Proposition Let M and M' be two motives in $\mathcal{M}_K^{av}(E)$, of rank 1 over E , such that

(i) M and M' become isomorphic over $\overline{K}$;

(ii) $p(M) = p(M')$ in $(K \otimes E \otimes \mathbb{C})^*/(E^*)^{Hom(K,\mathbb{C})}\ (K \otimes E)^*$.

Then $M \cong M'$ in $\mathcal{M}_K^{av}(E)$.

Proof. By I, 6.6.1, we have $M = M(\chi)$, $M' = M(\chi')$, for certain characters χ, χ' of K with values in E ; and (i) implies that $\chi' = \mu\chi$, for some finite order character of K - cf. ⓪, 3 and I, 6.15. Hence, by 1.8.3,

$$1 = \frac{p(M')}{p(M)} = p(\mu) .$$

By 3.2.3/4, this means that $\mu = 1$, and I, 5.1 finishes the proof. (In fact, a direct argument can be given, using (ii) once more.)

3.5.2 The proof of 3.5.1 shows that the $\overline{K}/K$-forms of rank-1-motives in $\mathcal{M}_K^{av}(E)$ are parametrized by the periods $p(\mu) \in (K \otimes E \otimes K^{ab})^*$, for μ running over the finite order characters of K .

4. The periods of Jacobi sum Hecke characters

4.0 The gamma function, i.e., the meromorphic continuation of

$$\Gamma(s) = \int_0^\infty e^{-x}x^s \frac{dx}{x} \qquad (\mathrm{Re}\ s > 0)$$

satisfies the following functional equations - for $s \in \mathbb{C}$, and $m \in \mathbb{Z}$, $m \geq 1$.

$$(4.0.0) \qquad s\Gamma(s) = \Gamma(1+s)$$

$$(4.0.1) \qquad \prod_{j=0}^{m-1} \Gamma\left(\frac{s+j}{m}\right) = (2\pi)^{\frac{m-1}{2}} m^{\frac{1}{2}-s} \Gamma(s)$$

$$(4.0.2) \qquad \Gamma(s)\Gamma(1-s) = \frac{\pi}{\sin \pi s} .$$

The first one implies that Γ induces a well-defined map

$$(4.0.3) \qquad \Gamma : \mathbb{Q}/\mathbb{Z} \to \mathbb{C}^*/\mathbb{Q}^* ,$$

and a folklore conjecture in transcendence theory says that all relations satisfied by (4.0.3), composed with

$$\mathbb{C}^*/\mathbb{Q}^* \to \mathbb{C}^*/\overline{\mathbb{Q}}^* ,$$

follow from (4.0.1) and (4.0.2) - see [LD], ex. 4.

4.1 The basic example

Let us resume the situation of I, 7.1.1, assuming $n \geq 3$. On the affine open part

$$Y_2^m + \ldots + Y_n^m = -1 \qquad (Y_i = \frac{X_i}{X_1})$$

of the Fermat hypersurface X_m^n, the n-2 form

$$Y_2^{\tilde{a}_2} \ldots Y_n^{\tilde{a}_n} \frac{dY_2}{Y_2} \wedge \ldots \wedge \frac{dY_{n-1}}{Y_{n-1}}$$

is an eigenform for the character $\underline{a} = \sum [a_j]$ of G_m^n, if $\frac{\tilde{a}_j}{m} (\mathrm{mod}\, \mathbb{Z}) = -a_j$, and $\tilde{a}_j \gg 0$. Its period against a suitable n-2 simplex is computed as

$$(4.1.0)\quad (2\pi i)^{-1} \prod_{j=1}^{n} (1-e^{2\pi i \cdot a_j}) \Gamma(-\langle a_j \rangle) ;$$

see [DMOS], I, 7.12-7.14. This allows us to compute the periods of the motive $M(\underline{a}) \times \mathbb{Q}(\mu_m)$ which, by construction, has the structure of a motive with coefficients in $\mathbb{Q}(\mu_m)$. But $M(\underline{a})$ is constructed in such a way that $M(\underline{a}) \otimes E$ is isomorphic, in $\mathcal{M}_{\mathbb{Q}}(\mathbb{Q}(\mu_m))$, to $R_{\mathbb{Q}(\mu_m)/\mathbb{Q}}(M(\underline{a}) \times \mathbb{Q}(\mu_m))$. Thus we find, using 1.3.1, 1.4.2 - 1.4.8, and, in the case where 0 is critical for $J(\underline{a})$, 1.7, ⓪, 8.2.7:

$$(4.1.1)\qquad p(M(\underline{a})) = \sqrt{d(\mathbb{Q}(\mu_m))} \cdot \prod_{\substack{k=1 \\ (k,m)=1}}^{m-1} \prod_{j=1}^{n} \Gamma(-a_j k)^{-1}$$

$$(4.1.2)\qquad c^+(M(\underline{a})) = \sqrt{d^+(\mathbb{Q}(\mu_m))} \cdot \prod_{\langle k\underline{a} \rangle > \langle -k\underline{a} \rangle} \prod_{j=1}^{n} \Gamma(a_j k)^{-1} ,$$

where $d(\mathbb{Q}(\mu_m))$ (resp. $d^+(\mathbb{Q}(\mu_m))$) is the discriminant of $\mathbb{Q}(\mu_m)$ (resp. of the maximal totally real subfield of $\mathbb{Q}(\mu_m)$ - see 1.7.12 (iv)). These expressions are well-determined up to a factor in $\mathbb{Q}^*$, as they should be for a motive in $\mathcal{M}_{\mathbb{Q}}(\mathbb{Q})$. Also, we claim that, if $s = 0$ is critical for $J(\underline{a})$, then

$$(4.1.3)\qquad c^-(M(\underline{a})) = \sqrt{d^-(\mathbb{Q}(\mu_m))} \cdot \prod_{\langle k\underline{a} \rangle > \langle -k\underline{a} \rangle} \prod_{j=1}^{n} \Gamma(a_j k)^{-1} ,$$

with $d(\mathbb{Q}(\mu_m)) = d^+(\mathbb{Q}(\mu_m)) \cdot d^-(\mathbb{Q}(\mu_m))$. In fact, by the behaviour of the discriminant in towers, we have

$$d^-(\mathbb{Q}(\mu_m)) = d^+(\mathbb{Q}(\mu_m)) \cdot [\prod_k (e^{2\pi i \frac{-k}{m}} - e^{2\pi i \frac{k}{m}})]^2$$

the product being over any set of representatives k of $(\mathbb{Z}/m\mathbb{Z})^*$ mod $\{\pm 1\}$. This shows that $(\sqrt{d^-(\mathbb{Q}(\mu_m))})_\tau \in (\mathbb{Q}(\mu_m) \otimes \mathbb{C})^*$ equals δ^- - given by 1.7.10 - , up to a factor in $\mathbb{Q}(\mu_m)^*$.

Note that - as it ought to be: [DP], 1.7 - $c^+(M(\underline{a}))$ and $i^{\varphi(m)/2}c^-(M(\underline{a}))$ are real, where φ is Euler's phi function.

4.2 Periods of Anderson's motives

If K is any abelian number field and $\underline{a} \in \mathbb{B}_K^0$ - see 0, 8.2 -, then the periods of Anderson's motive $M_K(\underline{a})$ - see I, 7.2; I, 7.4.5 - for the Jacobi sum character $J_K(\underline{a})$ can be computed by formulas which immediately generalize 4.1.1-4.1.3. In fact, note that periods are built into the notion of arithmetic Hodge structure - see I, 7.3.1/2. - We shall only give the final expression that Anderson obtains for the periods corresponding to the critical values of all Jacobi sum Hecke characters. It contains 4.1.2 as a special case, and 4.1.3 follows from it via 3.1.1. - The formula for $p(M_K(\underline{a}) \times K)$ is stated in 4.4.2.

4.2.1 If K is totally real, then F_∞ acts trivially on $H_B(M_K(\underline{a}))$, for any $\underline{a} \in \mathbb{B}_K^0$. (Essentially, this is so because $J_K(\underline{a})$ is "pulled down" from some totally imaginary extension of K.) Thus, by 2.0.2 above, the critical values of the character $J_K(\underline{a}) = \mu \cdot \mathbb{N}^{\sum n_a}$ - with $\underline{a} = \sum_a n_a[a]$ and μ of finite order - are just the elements of

$$\{s \in 2\mathbb{Z}+1 \mid s \leq \sum_a n_a\} \cup \{s \in 2\mathbb{Z} \mid s > \sum_a n_a\} .$$

We put

$$\mathrm{Crit}_K(\underline{a}) = 2\mathbb{Z} \cap (\sum_a n_a, \infty) .$$

4.2.2 If K is totally imaginary, then 2.0.3 implies that the critical s for $J_K(\underline{a})$ are precisely those in

$$\mathrm{Crit}_K(\underline{a}) = \left\{ s \in \mathbb{Z} \;\middle|\; \begin{array}{l} \langle t\underline{a}\rangle < s \leq \langle t\, c\, \underline{a}\rangle \text{ for all } t \in G(\overline{\mathbb{Q}}/\mathbb{Q}) \\ \text{with } \langle t\underline{a}\rangle \leq \langle t\, c\, \underline{a}\rangle \end{array} \right\}$$

Here, as usual, c denotes complex conjugation; the galois

action is that defined in $\mathbb{O}$, 8.2.1.

4.2.3 Notation. $d(K)$, resp. $d^+(K)$, denotes the discriminant of K , resp. of the maximal totally real subfield of K ; and $d(K) = d^+(K) \cdot d^-(K)$. For all $\underline{a} \in \mathbb{B}$, $\underline{a} = \sum_a n_a[a]$, extend 4.0.3 by the rule

$$\Gamma(\underline{a}) = \prod_a \Gamma(a)^{n_a} \in \mathbb{C}^*/\mathbb{Q}^* .$$

4.2.4 For all abelian number fields K , all $\underline{a} \in \mathbb{B}_K^0$ and all $s = n \in \mathrm{Crit}_K(\underline{a})$, one finds

$$(4.2.5) \quad c^+(M_K(\underline{a})(n)) = \pi^{n[K:\mathbb{Q}]/2} |d^{(-)^n}(K)|^{1/2} \prod_{\substack{\sigma \in G(K/\mathbb{Q}) \\ \langle \sigma\underline{a}\rangle \geq \langle \sigma c \underline{a}\rangle}} \Gamma(\sigma \underline{a})^{-1} .$$

4.3 Lichtenbaum's "Γ-hypothesis"

As Anderson points out, the period calculation 4.2.5, joined with theorem 2.1 above, yields the following theorem on the critical L-values of Jacobi sum Hecke characters which contains the most general formulation of what Lichtenbaum had called his Γ-hypothesis - see [Li], [KL].

4.3.1 Theorem. For all abelian number fields K , all $\underline{a} \in \mathbb{B}_K^0$, and all $s = n \in \mathrm{Crit}_K(\underline{a})$,

$$\pi^{-n[K:\mathbb{Q}]} |d^{(-)^n}(K)|^{1/2} \prod_{\langle \sigma\underline{a}\rangle \geq \langle \sigma c \underline{a}\rangle} \Gamma(\sigma\underline{a}) \cdot L(J_K(\underline{a}),n) \in \mathbb{Q} .$$

Note that, in deriving this statement from 2.1, one has to use, as in 4.1, that $M_K(\underline{a}) \otimes K$ is isomorphic, in $\mathcal{M}_{\mathbb{Q}}(K)$, to $R_{K/\mathbb{Q}}(M_K(\underline{a}) \times K)$, where $M_K(\underline{a}) \times K$ has a natural structure of a motive with coefficients in K that makes it into a motive for $J_K(\underline{a})$. Recall also that $J_K(\underline{a})$ is galois equivariant - see $\mathbb{O}$, 8.2.5 - , so that the L-functions of all of its conjugates coincide.

Finally, it should be pointed out that, in the case that K is totally real, every critical s for $J_K(\underline{a})$ either lies in $\mathrm{Crit}_K(\underline{a})$ or is related to an element of $\mathrm{Crit}_K(-\underline{a})$, by the functional equation.

4.4 Γ-relations

4.4.1 Theorem. *Let $K \subset \overline{\mathbb{Q}}$ be an abelian number field. If $\underline{a},\underline{b} \in \mathbb{B}_K^0$ satisfy $J_K(\underline{a}) = J_K(\underline{b})$, then*

$$(\Gamma(\sigma\underline{a}))_{\sigma \in G(K/\mathbb{Q})} = (\Gamma(\sigma\underline{b}))_{\sigma \in G(K/\mathbb{Q})} ,$$

in $(K \otimes \mathbb{C})^*/K^*$.

Proof. By construction, the motive $M_K(\underline{a}) \times K$ has a natural structure of a motive with coefficients in K with respect to which it is a motive for $J_K(\underline{a})$. For all σ, one has

(4.4.2) $$p(M_K(\underline{a}) \times K; \mathrm{id}_K, \sigma) = \Gamma(\sigma\, c\, \underline{a})^{-1} .$$

(The fact that complex conjugation creeps into this formula is clearly seen in our basic example: I, 7.1./2, and 4.1.0 above.) By I, 5.1, the theorem follows.

q.e.d.

4.4.3 Corollary ([A2], 8.6) *If $\underline{a}$ and $\underline{b}$ are in B_K^0 such that $\mathrm{Crit}_K(\underline{a}) \neq \emptyset$, and*

$$L(J_K(\underline{a}),s) = L(J_K(\underline{b}),s) ,$$

as meromorphic functions on $\mathbb{C}$, then

$$\prod_{\langle\sigma\underline{a}\rangle \geq \langle\sigma c \underline{a}\rangle} \Gamma(\sigma\underline{a}) = \prod_{\langle\sigma\underline{b}\rangle \geq \langle\sigma c \underline{b}\rangle} \Gamma(\sigma\underline{b}) ,$$

in $\mathbb{C}^*/\mathbb{Q}^*$.

Proof. The hypothesis implies: immediately, that $\mathrm{Crit}_K(\underline{a}) = \mathrm{Crit}_K(\underline{b})$; and, modulo an exercise in analytic number theory, that $J_K(\underline{b}) = J_K(\tau\underline{a})$, for some $\tau \in G(K/\mathbb{Q})$. Then, the theorem yields what is claimed, in view of 0, 8.2.7.

Known variants of the theorem used to be encouraging companions to the Γ-hypothesis when this was still unproven. Its motivic proof is a nice illustration of our central theme: how to derive period relations from character identities. More precisely, it is a compatibility result inside one family of motives for a class of Hecke characters. In that sense it is the analogue, for Anderson's motives, of Shimura's monomial relations, as derived from the standard motives of Hecke characters in chapter IV below.

4.4.4 A different instance of our main theme occurs when $J_K(\underline{a})$ is of finite order, and $M_K(\underline{a})$ is compared to an Artin motive. This was already pointed out by Deligne in [DP], 8.9 - 8.13. Let us briefly rederive the results in our setting.

By 0, 8.2.7, the Jacobi sum Hecke character $J_K(\underline{a})$ is of finite order if and only if $\langle\sigma\underline{a}\rangle = 0$, for all $\sigma \in \mathrm{Gal}(K/\mathbb{Q})$. If this is so, then - by I, 5.1 - $M_K(c\underline{a}) \times K$ is isomorphic, in $\mathcal{M}_K(K)$, to the Artin motive of $J_K(c\underline{a}) = J_K(\underline{a})^{-1}$, and we deduce from 4.4.2, for $M_K(c\underline{a})$, and 3.2.4, with $\tilde{\sigma} = \tau = \mathrm{id}$, the following theorem which contains conjecture 8.13 of [DP], and, together with 4.4.6, is equivalent to theorem 7.18 in [DMOS], chap. I. It also implies, of course, 4.4.1 and 4.4.3 above.

4.4.5 Theorem. For all abelian number fields K , and all $\underline{a} \in \mathbb{B}_K^0$ such that $\langle\sigma\underline{a}\rangle = 0$ for each $\sigma \in G(K/\mathbb{Q})$, reading the finite order character $J_K(\underline{a})$ on $\mathrm{Gal}(\overline{\mathbb{Q}}/K)^{ab}$, one has

(i) $\Gamma(\underline{a}) \in \overline{\mathbb{Q}}^*/\mathbb{Q}^*$

(ii) $\Gamma(\underline{a})^s = J_K(\underline{a})(s) \cdot \Gamma(\underline{a})$, for all $s \in G(\overline{\mathbb{Q}}/K)$.

Part (i) was first proved directly by Koblitz and Ogus in the appendix to [DP]. - It is shown in [Sch Γ] that a good deal of

(ii) can be derived from 4.0.1, 4.0.2 using only classical results on the arithmetic of Gauss sums.

4.4.6 As in [DMOS], I, 7.18, the preceding theorem can be complemented to give the behaviour of $\Gamma(\underline{a})$ under all of $\mathrm{Gal}(\overline{\mathbb{Q}}/\mathbb{Q})$:

(iii) For all $t \in G(\overline{\mathbb{Q}}/\mathbb{Q})$, the quotient $\frac{\Gamma(\underline{a})}{\Gamma(\underline{a})^t}$ lies in K^*, and for all $\sigma \in \mathrm{Gal}(K/\mathbb{Q})$, one has

$$\left(\frac{\Gamma(\underline{a})}{\Gamma(\underline{a})^t}\right)^{\sigma} = \frac{\Gamma(\sigma\underline{a})}{\Gamma(\sigma\underline{a})^t} .$$

Proof. Writing ξ the "period" 3.2.3 of the Artin motive of $J_K(\underline{a})$, we have

$$\Gamma(\underline{a})^t = \xi^{t \otimes t} ,$$

for all $t \in G(\overline{\mathbb{Q}}/\mathbb{Q})$. This and the galois equivariance ①, 8.2.5:

$$J_K(\underline{a})(s) = J_K(t\underline{a})(t^{-1}st) \quad (s \in G(\overline{\mathbb{Q}}/K))$$

easily imply that $\Gamma(\underline{a})/\Gamma(\underline{a})^t \in K^*$.

The last claim is proved by analysing the action of $\mathrm{Gal}(\overline{\mathbb{Q}}/\mathbb{Q})$ on $\mathrm{End}_{/\overline{\mathbb{Q}}}(M_K(\underline{a})) = K$: just imitate the argument on [DMOS], p. 93; the details are left to the reader.

CHAPTER THREE:

Elliptic Integrals and the Gamma Function

The subject of this chapter is a natural continuation of II, 4.4: - we now compare Anderson's motives for Jacobi sum Hecke characters to elliptic curves with complex multiplication. This gives essentially a refinement of the so-called formula of Chowla and Selberg - which originally is due to M. Lerch.

1. A formula of Lerch

1.0 Let $K \overset{1}{\hookrightarrow} \mathbb{C}$ be an embedded imaginary quadratic field, $-D$ its discriminant, and $J_D = J_K(\underline{a}_D)$ the basic Jacobi sum Hecke character of K defined in 0, 8.1 for $D \neq 3, 4, 8$ and in 0, 8.3 for arbitrary D. The infinity type T_D of J_D (0, 8.1.5) is written

$$T_D = n_1 \cdot 1 + n_c \cdot c \quad ,$$

and h_D denotes the class number of K.

Let χ be any fixed Hecke character of K (with values in some CM field $E \supset K$) whose infinity type is -1. Then there exists a character of finite order μ of K, with values in E, such that

$$(1.1) \qquad \mu \cdot \chi^{h_D} = J_D^{-1} \cdot \mathbb{N}^{n_c} \quad .$$

Let us now compute the periods c^+ of motives attached to both sides of the equation. They have to be equal by I, 5.1. On the left hand side, use II, 3.3.3 for the present μ, and II, 1.8.1/3 as well as II, 1.7.12(iv). On the right, use II, 4.2.5 observing 0, 8.1.3 and 0, 8.3.1, and remembering that $(M_K(-\underline{a}_D) \times K) \otimes E$ is a motive for J_D^{-1} considered as Hecke character of K with values in E. This gives, for $D \neq 3,4,8$:

$$(1.2) \qquad \left(\tilde{\eta}_{\tau,\tau} \cdot p(\chi;\tau|_K,\tau)^{h_D}\right)_\tau = \left(\pi^{-n_c} \, |d^{(-)^{n_c}}(K)|^{1/2} \prod_{\substack{j=1 \\ \epsilon(j)=1}}^{D-1} \Gamma(\tfrac{j}{D})\right)_\tau$$

where τ runs over the complex embeddings of E, and ϵ is the

Dirichlet character corresponding to $K \subset \mathbb{Q}(\mu_D)$.

Let F be any finite abelian extension of K such that χ takes values in K^* on ideals which are norms from F . (Note that F has to contain the Hilbert class field H of K .) Then there exists an elliptic curve A defined over F such that $H_1(A)$ is a motive for the Hecke character $\psi = (\chi \circ N_{F/K}$, considered as character of F with values in K). This is a special case of Casselman's theorem, i. e., theorem 6 in [Shi L] - cf. I, 4.1.1. Note that, in terms of A , the inclusion $K \hookrightarrow F$ is given by the action of $\mathrm{End}\, A$ on the tangent space of A at the origin. - Cf. also [GS] § 4 .

From II, 1.8.6 and II, 1.8.8, we find that, for all $\tau : E \hookrightarrow \mathbb{C}$ which restrict to $\tau_o : K \to \mathbb{C}$,

$$(1.2.1) \qquad p(\chi;\tau_o,\tau)^{h_D\cdot[F:H]} = \prod_{\sigma|_K = \tau_o} p(\psi;\sigma,\tau_o) ,$$

up to a factor in E^* .

<u>1.2.2</u> By II, 1.5.1 and II, 1.6.6, $p(\psi;\sigma,\tau_o)$ is, independently of τ_o and σ with $\sigma|_K = \tau_o$, equal (up to the usual indeterminacy) to Ω_σ , a fundamental period of the elliptic curve A^σ/F^σ . In other words, the complex lattice Λ_σ corresponding to the pair $(A^\sigma(\mathbb{C}),\omega^\sigma)$, for a holomorphic 1-form ω on A/F whose class is an $F \otimes K$ basis of $H^1_{DR}(A)$, satisfies

$$\Lambda_\sigma \cdot \mathbb{Q} = \Omega_\sigma \cdot K \subset \mathbb{C} .$$

We now make the following

<u>1.3 Assumption.</u> F may be chosen to be the Hilbert class field H of K . In other words, χ takes values in K^* on all principal ideals on which it is defined.

<u>1.3.1. Remark.</u> Characters χ of type -1 which satisfy 1.3 exist for all imaginary quadratic fields K - their construction is straightforward. The field of values E is then of degree h_D over K - see [Ro], cf. [Sch O], E . - It can have subfields which are galois over K only insofar as the few roots of unity in K^* afford Kummer extensions corresponding to elements in the class

group of K - cf. [Gr 1], § 15 for the case where h_D is odd (<u>and</u> ψ equivariant under complex conjugation).

<u>1.3.2.</u> Since J_D is galois equivariant, and therefore, in particular, takes values in K, and since h_D kills the class group of K, 1.1 and 1.3 imply that μ takes values in K. Thus, <u>if</u> $D \neq 3,4$, <u>then</u> μ <u>is at most quadratic.</u> In this case, the factor $\tilde{\eta}_{\tau,\tau}$ simply becomes, independently of τ, any non-zero element Δ_μ of $K^{ab} \subset \bar{\mathbb{Q}}$ such that

$$\Delta_\mu^s = \mu(s)\Delta_\mu , \quad \text{for all } s \in \mathrm{Gal}(\bar{\mathbb{Q}}/K) .$$

By 1.2.1, 1.2.2, and 1.3.2, formula 1.2 becomes an identity of vectors with identical components:

<u>If</u> $D \neq 3,4,8$, <u>then, up to a factor in</u> K^*,

$$(1.4) \qquad \Delta_\mu \cdot \prod_{\sigma \in G(H/K)} \Omega_\sigma \underset{K^*}{\sim} \left(\frac{\sqrt{D}}{\pi}\right)^{\frac{1}{2}\left(\frac{\varphi(D)}{2} - h_D\right)} \prod_{\substack{j=1 \\ \varepsilon(j)=1}}^{D-1} \Gamma\left(\tfrac{j}{D}\right)$$

<u>1.4.1</u> Before discussing 1.4 let us write down the corresponding relations for $D = 3,4,8$. In these cases we simply take $\chi = J_D^{-1}$ in 1.1: all three class numbers are 1. The corresponding elliptic curves A_D/K were briefly discussed in I, 7.5. They are actually defined over $\mathbb{Q}$ and we have isomorphisms of motives in $\mathcal{M}_{\mathbb{Q}}(\mathbb{Q})$:

$$H^1(A_D) \cong M_K(\underline{a}_D) ,$$

because A_D is constructed such that $L(H^1(A)/\mathbb{Q},s) = L(J_D,s)$. Thus, writing $\Omega_D = \int_{A_D(\mathbb{R})} \omega$ the real period of a nonzero differential of the first kind ω on $A_D/\mathbb{Q}$, we find the following identities of classes in $\mathbb{C}^*/\mathbb{Q}^*$:

$$\Omega_3 \underset{\mathbb{Q}^*}{\sim} \Gamma(c\underline{a}_3) = \frac{\Gamma(\frac{1}{3})^2}{\Gamma(\frac{2}{3})} .$$

$$\Omega_4 \underset{\mathbb{Q}^*}{\sim} \Gamma(c\underline{a}_4) = \frac{\Gamma(\frac{1}{2})\Gamma(\frac{1}{4})}{\Gamma(\frac{3}{4})}$$

$$\text{(by II, 4.0.2)} \quad = \quad \frac{\Gamma(\frac{1}{4})^2}{\sqrt{2\pi}} \quad .$$

$$\Omega_8 \underset{\mathbb{Q}^*}{\sim} \Gamma(c\underline{a}_8) \quad = \quad \frac{\Gamma(\frac{1}{8})\Gamma(\frac{3}{8})}{\sqrt{\pi}} \quad .$$

To be sure, in the first two cases, it seems much more natural to go the other way: the formulas for Ω_3 and Ω_4 are classically well-known (cf. § 2 below), and they imply, by II, 3.5, that $A_D \cong M_K(\underline{a}_D)$ <u>over</u> K , and in fact over $\mathbb{Q}$. This then shows what was claimed in I, 7.5: that, for $D = 3,4$, the elliptic curves A_D described are such that $H^1(A_D)$ is a motive for J_D .

<u>1.4.2</u> Multiplying 1.4 with its complex conjugate yields a relation up to a rational number. In order to put it into a classical shape we apply II, 4.0.2 to the product on the right once, and use the following relation which is proved by arguments of the kind well-known in the context of the analytic class number formula for real quadratic fields:

$$(1.4.3) \qquad \prod_{\substack{j=1 \\ \epsilon(j)=1}}^{D} \sin(\pi \tfrac{j}{D}) \underset{\mathbb{Q}^*}{\sim} \sqrt{D}^{h_D} \; .$$

This version holds for <u>all</u> $D > 0$ such that $-D$ is the discriminant of a quadratic field. In fact, the more natural right hand side, $\sqrt{D}^{\varphi(D)/2}$ was replaced by $\sqrt{D}^{h_D}$ in order to make it come out right for $D = 8$.

Thus, writing $2m$ the number of units of K , we get the following relation, which can be checked for the exceptional cases $D = 3,4,8$ directly from 1.4.1:

$$(1.4.4) \qquad \Delta_\mu \bar{\Delta}_\mu \prod_{\sigma \in G(H/K)} [\tfrac{\sqrt{D}}{\pi} \Omega_\sigma \bar{\Omega}_\sigma] \underset{\mathbb{Q}^*}{\sim} \prod_{(\mathbb{Z}/D\mathbb{Z})^*} \Gamma(\tfrac{j}{D})^{\epsilon(j)m} \; .$$

Note that the complex conjugate $\bar{\Delta}_\mu$ of Δ_μ is not intrinsically defined: $K(\Delta_\mu)$ need not be a CM field. But, as $\Delta_\mu^2 \in K^*$, $\bar{\Delta}_\mu$ is well-determined up to a sign - which is inessential for 1.4.4.

1.4.5 Up to the interpretation of the factor $\Delta_\mu \overline{\Delta}_\mu \sqrt{D}^{\,h_D}$, 1.4.4 is easily seen to be the exponential of a precise identity found by M. Lerch in 1897 (and rediscovered later by Chowla and Selberg), taken modulo $\mathbb{Q}^*$ - see § 2 below. In this analytic identity, the factor $\Delta_\mu \overline{\Delta}_\mu \sqrt{D}^{\,h_D}$ appears as the 12-th root of $\prod_\sigma \Delta(\Lambda_\sigma)$, where $\Delta(\Lambda_\sigma)$ is the discriminant of the lattice Λ_σ mentioned in 1.2.2. - See 1.5.6 below.

1.5.1 The left hand side of 1.4 - or of 1.4.4 - only depends on the field K. In fact, two elliptic curves over H coming from different characters χ of K (like in 1.2) are twists of each other, by a finite order character of H of the form $\mu \circ N_{H/K}$ - so we can use II § 3. Similarly, if any elliptic curve C/H with complex multiplication by K is given, it will be the twist of an A like in 1.2, by a finite order character of H - and again II § 3 tells us by which factor in H^{ab} to modify the left hand side of 1.4 in order to get the formula for the product of periods of the C^c. - For the more general case where C is defined over some $F \supset H$, see § 3 below. There we shall also discuss a possible motivic interpretation of 1.4.4.

1.5.2 H/K-curves.

An elliptic curve A with complex multiplication by K defined over H is called an H/K-curve, if it is H-isogenous to all conjugates A^σ, with $\sigma \in G(H/K)$. If ψ is the Hecke character of H with values in K such that $H_1(A) = M(\psi)$, then A is an H/K-curve if and only if, for all $\sigma \in G(H/K)$ and all ideals $\mathfrak{a}$ of H on which ψ is defined, one has

$$\psi(\mathfrak{a}^\sigma) = \psi(\mathfrak{a}) .$$

If A is an H/K-curve, then

$$\psi^{h_D} = \psi \circ i \circ N_{H/K} ,$$

where i is the inclusion of ideals of K into ideals of H. The character $\psi \circ i$ of K with values in K satisfies

(1.5.3) $$\mu\cdot(\psi\circ i) = J_D^{-1}\cdot \mathbb{N}^{n_c} ,$$

for some character μ of K of exponent two. Now we can imitate the arguments that have led us to 1.2, and from there to 1.4, obtaining 1.4 with the periods Ω_σ of the A^σ . It is not always easy, however, to identify the character μ for a given H/K-curve A .

1.5.4 Standard $\mathbb{Q}$-curves

Assume $D > 3$ is odd, and recall from [Gr 1] § 11 the fundamental Hecke character of H attached to the field $K = \mathbb{Q}(\sqrt{-D})$: reading the Dirichlet character ϵ of K as a character

$$\epsilon : \mathfrak{o}_K/\sqrt{-D}\cdot\mathfrak{o}_K \rightarrow \{\pm 1\} ,$$

every principal ideal of K prime to $\sqrt{-D}$ admits an unique generator $\alpha \in K^*$ with $\epsilon(\alpha) = 1$. Call χ_D any extension to all ideals of K prime to D , so χ_D is a Hecke character of K with values in some CM field E of degree h_D over K . Put $\psi_D = \chi_D\circ N_{H/K}$ - this is the fundamental Hecke character of H with values in K we were alluding to above. We claim that

(1.5.5) $$\chi_D^{h_D} = J_D\,\mathbb{N}^{-n_c} ,$$

with no twisting character μ .

One way to prove (1.5.4) is by direct attack. - We leave this as an exercise to the reader, noting only that, if h_D is odd, one can get away without really looking at the definition of J_D - see [BL], lemma 3.4. - This direct proof of 1.5.4 verifies 1.4 with $\Delta_\mu = 1$, and Ω_σ the period of A_D^σ , with A_D/H being the standard $\mathbb{Q}$-curve of K : $H^1(A_D) = M(\psi_D)$.

Alternatively, one can show that $\mu = 1$ if one has an independent way of checking that 1.4 holds for the Ω_σ of A_D , with $\Delta_\mu = 1$ - cf. II, 3.5. Now, Gross has shown - see [Gr 1], 21.2.2 and [Gr 3], 5.6 - that the analytically proved relation 1.4.4 can be refined directly to yield exactly this: 1.4 for A_D with $\Delta_\mu = 1$. (In the case h_D odd, this elementary argument of Gross was quite an important ingredient in the proof of the Γ-hypothesis for imaginary quadratic fields of odd class number: see [BL], a paper completed before the advent of Anderson's motives.)

1.5.6 For any D, let A/H and A'/H be elliptic curves with complex multiplication by K, with characters ψ, ψ', respectively, so that

$$H_1(A) = M(\psi) \quad \text{and} \quad H_1(A') = M(\psi') .$$

Put $\nu = \psi'/\psi$. Assume we know Δ_μ in 1.4 applied to the periods Ω_σ of A^σ. Then, for Ω'_σ corresponding to A'^σ, one has

$$\prod_\sigma \Omega'_\sigma \underset{K^*}{\sim} \Delta_{(\nu\circ i)} \cdot \prod_\sigma \Omega_\sigma , \tag{1.5.7}$$

where i is as in 1.5.3. This then determines the factor Δ'_μ which has to be used in 1.4 for the Ω'_σ. - Cf. [BG], 10.5, where the analysis of the factors is finer than our motivic methods permit. There, Gross refers back to [Gr 4] § 4; cf. the analogous passages in [GS], §§ 4, 9. - The formula recalled after [GS], 10.1, implies that the two expressions $\Delta_\mu \overline{\Delta}_\mu \sqrt{-D}^{\,h_D}$ and $\prod_\sigma \Delta(\Lambda_\sigma)$ - see 1.4.5 - do transform the same way under changement of the curve A/H. So, it is enough to check they are equal for one such A. This is what we have indicated, in 1.5.4, for D odd. - For curves that are no longer defined over H, see 3.2 below.

2. An historical aside

I am indebted to R. Sczech for pointing out to me that an analytic formula which implies 1.4.4 occurs as identity no. 163 in E. Landau's paper [La]. Thanks to Landau's bibliographical scrutiny, this paper contains references to what probably is a fairly complete history of this formula, prior to 1903.

Special cases, including $\mathbb{Q}(\sqrt{-4})$ - the lemniscatic case - and $\mathbb{Q}(\sqrt{-3})$ were known early in the 19th century, the main reference being Legendre's book [Le] - e.g., 1ère partie, n° 146, 147; pp. 209 f. By the middle of that century, the lemniscatic formula of 1.4.1 above could be used, without further comment, by Eisenstein: [Ei 1], p. 186. - It is this part of the history that Chowla and Selberg were aware of when writing their papers: the announcement [CS] (see § 4 for our formula) and the final version [SC] (§§ 8, 12). - Cf. [WW], 22.8.

In the analytic proofs of 1.4.4, the Γ-values, or rather their logarithms, usually enter through the evaluation of

$$L'(\varepsilon,1) = -\sum_{n=1}^{\infty} \varepsilon(n) \frac{\log n}{n} ,$$

via Kummer's series for $\log \Gamma(x)$ - i.e., the identity derived in [Ku]. This part seems to have been done first by A. Berger - see [Be], p. 29/30 - as early as 1883. When Lerch rediscovered this argument in 1897 - [Ler], p. 302 f - Kronecker, using his "first" limit formula - cf. [We], VIII § 6 - , had already expressed $L'(\varepsilon,1)$ in terms of various constants <u>and</u> (logarithms of) special values of theta series which correspond to the Ω_σ's in our notation - see [Kr], art. XVI, formula 7. Putting both parts together, Lerch deduces our identity (more precisely, its logarithm) as formula 26 of [Ler], p. 303.

Weil points out - [We 2], IX §§ 2,4 - that Lerch could have used his determination, in 1894, of the derivative at $s=0$ of the Hurwitz zeta function, in order to relate $L'(\varepsilon,1)$ to values of $\log \Gamma$. But using Kummer's series for this seems to have been closer to the taste of the day: in fact, J. de Séguier, a Jesuite professor of mathematics at the University of Angers, rediscovered this part of the proof in 1899 - see [dS 2] § 10 - although, as Landau does not fail to point out ([La], p. 177), he looses a factor of $\frac{\pi}{2}$ along the way. de Séguier should have been especially well prepared to put together both parts of the proof because he had published, in 1894, a whole book - [dS 1] - on Kronecker's series of mémoires [Kr]. But our identity does not seem to have caught his interest.

Landau, in [La], gives new proofs of both parts of our identity. The part done first by Berger allows him to illustrate the use of divergent Dirichlet series - the theory of which he develops in the 3rd section of this paper -, see [La], 177-179.

I do not know of any occurence of our identity between [La] and [CS], i.e., between 1903 and 1949.

Weil seems to have been the first to envisage a geometric proof of our identity - see [W III], 1976 b (and 1976 b*). He did not succeed in general; but in a slightly later manuscript - [W III], 1977 c - he provided what Gross could then develop into an essential tool for his geometric proof of 1.4 <u>up to factors</u> in $\overline{\mathbb{Q}}^*$ - see [Gr 2]. This proof tracks down periods along a family of abelian varieties which, at one point, contains (a factor of) the Jacobian of a Fermat curve - whence the Γ-values -, at another a power of the elliptic curve A - whence $\Omega \underset{\overline{\mathbb{Q}}^*}{\sim}$ any Ω_σ. Gross' deformation argument, in turn, provided a key step in Deligne's proof of the fundamental theorem I, 2.1.3 on absolute Hodge cycles on abelian varieties - see [DMOS], I, 4.8, 4.11 - : it enabled him to show that certain "exceptional cycles", presented in [W III], 1977 c as would-be obstructions to the Hodge conjecture, were at least absolutely Hodge. Since it is basically this theorem I, 2.1.3 - along with Anderson's motives - on which our deduction of 1.4 rests, the story of the geometric proofs has come almost full circle.

An amazing kind of a geometric revindication of the log in front of 1.4.4 which naturally comes out of the analytic proofs is provided by Falting's theory of the modular height of abelian varieties - cf. [DF], 1.5. Should the original identity really be viewed as an identity of the (logarithmic) heights of two ... ?

The attempt, in [Mor], to generalize the identity along analytic lines has, so far, not been linked to the geometric vein. The same can be said of the analogues for real quadratic fields in [Den]. If these theories have a geometric meaning it can be expected to be fairly different from the one we encountered with 1.4.

3. Twists and multiples

<u>3.0</u> Either one of the following two properties characterize the imaginary quadratic fields among all CM fields.

3.0.1 The set of all CM types of K forms a $\mathbb{Z}$ basis of the group of infinity types of all Hecke characters of K ;

3.0.2 Each element of $\mathbb{Z}[\mathrm{Hom}(K,\mathbb{C})]$ is the infinity type of a Hecke character of K .

Taking everything modulo $\sum \sigma$, the infinity type of the norm $\mathbb{N}$ on K , i.e., $1+c$ if K is imaginary quadratic, these conditions amount to:

3.0.3 The group of all infinity types of Hecke chracters of K , taken modulo $\sum \sigma$, is a free $\mathbb{Z}$ module of dimension 1.

We know, from ◐, 8.4.3, that the subgroup St_K of infinity types of Jacobi sum Hecke characters of K , taken modulo its element $1+c$, is precisely $h\,.(\mathbb{Z}[\mathrm{Hom}(K,\mathbb{C})]/\sum \sigma)$, for h the class number of the imaginary quadratic field K . The refinement 1.4 of Lerch's period relation 1.4.4 was deduced by writing the generator T_D of $St_K/\sum \sigma$ as $h\cdot\alpha \in h\cdot(\mathbb{Z}[\mathrm{Hom}(K,\mathbb{C})]/\sum \sigma)$. Since we are working in a onedimensional $\mathbb{Z}$ module, the following remark is plain.

3.1 Remark. If, in the arguments 1.1 - 1.4, the character J_D is replaced by any Jacobi sum Hecke character of K (and n_c , h_D, μ are changed accordingly), then the period relations between elliptic integrals of CM type and values in $\Gamma(\mathbb{Q})$ that one finds are all powers of 1.4, up to twisting by the norm or by finite order characters - see II § 3 - , and up to Γ-relations - which should all follow from II, 4.0.

3.1.1 In [St], Stern, proving a conjecture of Legendre, shows that at most $\frac{\varphi(D)}{2}$ of the values $\{\Gamma(\frac{j}{D}) \mid 0 < j < D\}$ are independent with respect to the relations II, 4.0.1/2. Unlike Landau - [La], p. 179 - we do not see 1.4 (or 1.4.4) as a relation which "allows to reduce the number of independent values" in this set. Instead, 1.4 goes beyond II, 4.0 in that it relates two different kinds of transcendental constants: elliptic integrals and Γ-values. This is also the use which is made of 1.4 in transcendence theory: to transport transcendence results from elliptic integrals to certain combinations of Γ-values ...

A remark similar to 3.1 also applies if we look at 1.4 from the point of view of the elliptic curves:

3.2 Let F be a finite extension of H, and A'/F an elliptic curve with complex multiplication by K. Write ψ' the Hecke character of F with values in K such that $H_1(A') = M(\psi')$. As in 1.5.6, let us compare A' to a curve A/H, $H_1(A) = M(\psi)$, for which 1.4, with all of its constants, is assumed to be known. Then $\psi' = \nu\cdot(\psi \circ N_{F/H})$, for some finite order character ν of F, with values in K. It follows, as in 1.5.7, that (assuming F/K galois, for simplicity)

$$(3.2.1) \qquad \prod_{\sigma \in G(F/K)} \Omega'_\sigma \underset{K^*}{\sim} \Delta(\nu\circ i) \cdot \Big(\prod_{\sigma \in G(H/K)} \Omega_\sigma\Big)^{[F:H]} ,$$

where i is the inclusion of ideals of K into ideals of F. Then 1.4 yields, for $D \neq 3,4,8$,

$$(3.2.2) \qquad \frac{\Delta_\mu^{[F:H]}}{\Delta(\nu\circ i)} \prod_{\sigma \in G(F/K)} \Omega'_\sigma \underset{K^*}{\sim} \Big\{\Big(\frac{\sqrt{D}}{\pi}\Big)^{\frac{1}{2}\left(\frac{\varphi(D)}{2} - h_D\right)} \cdot \prod_{\substack{j=1 \\ \varepsilon(j)=1}}^{D-1} \Gamma\Big(\frac{j}{D}\Big)\Big\}^{[F:H]} .$$

3.3 Gross once asked me for a direct motivic interpretation of Lerch's relation 1.4.4. Remember that we have obtained 1.4.4 by mindlessly multiplying 1.4 with its complex conjugate. - I propose the following identity of periods in $\mathbb{C}^*/\mathbb{Q}^*$ as an answer to this question:

$$(3.3.1) \qquad c^+([R_{K/\mathbb{Q}} M(\mu\cdot\chi^{h_D})]|_{\mathbb{Q}}) = c^+(M(\underline{a}_D)(-n_c) \otimes K|_{\mathbb{Q}}) .$$

In fact, 1.4.4 is deduced from 3.3.1, using the arguments of 1.2 - 1.4 as above, by virtue of II, 1.8.2, an analogue of II, 1.3.2 for c^+, and II, 1.6.6.

Note that, in case $h_D = 1$, 3.3.1 corresponds to studying the Hasse-Weil L-function of the elliptic curve A, instead of separating its CM factors $L(\chi,s)$ and $L(\overline{\chi},s)$.

CHAPTER FOUR:

Abelian Integrals with Complex Multiplication

In III we have studied the relations between periods of Hecke characters of imaginary quadratic fields and values of the gamma function. One aim of this chapter is to generalize these results to Hecke characters of abelian CM fields - see § 2. In order to do so, however, we first have to analyze a phenomenon which occurs for all CM fields K of degree $[K : \mathbb{Q}] > 2$: the monomial period relations implied by $\mathbb{Z}$ linear relations among CM types of K. These relations were discovered by Shimura - see [Shi P], [Shi O] - ; their motivic version (up to factors in $\overline{\mathbb{Q}}^*$) is already present in [DP], 8.18 - 8.23; and their motivic proof (up to $\overline{\mathbb{Q}}^*$) was explained in [DB] .

1. Shimura's monomial relations

1.0 Let K be an algebraic number field, and χ_i , for $i=1,\dots,r$, a collection of algebraic Hecke characters of K all of which take values in one number field E . Assume there are integers n_i such that

(1.0.0) $$\prod_{i=1}^{r} \chi_i^{n_i} = \mu ,$$

for some character of finite order μ . Then, by II, 1.8.3, we get the period relation

(1.0.1) $$\prod_{i=1}^{r} p(\chi_i)^{n_i} = p(\mu) \quad \text{in} \quad (K \otimes E \otimes \mathbb{C})^*/(E^*)^{\mathrm{Hom}(K,\mathbb{C})} \cdot (K \otimes E)^* .$$

Furthermore, we know how to compute $p(\mu)$ from μ : - see II, 3.2. In particular, for all $\sigma \in \mathrm{Hom}(K,\mathbb{C})$, $\tau \in \mathrm{Hom}(E,\mathbb{C})$, the complex number $p(\mu;\sigma,\tau)$ lies in the composite of the maximal abelian extension of K^σ with E^τ ;

(1.0.2) $$p(\mu;\sigma,\tau) \in [(K^\sigma)^{ab} \cdot E^\tau]^* \subset \mathbb{C}^* .$$

1.1 Shimura's basic relations

1.1.0 Assume that, in the situation of 1.0, each χ_i is of weight -1 , and that, for all $\sigma \in \mathrm{Hom}(K,\mathbb{C})$, $\tau \in \mathrm{Hom}(E,\mathbb{C})$, the Hodge exponents $n_i(\sigma,\tau)$ of χ_i - see ●, § 4 - are all either -1 or 0 , for all $i = 1,\dots,r$. Then 1.0.0, with unspecified μ of finite order, is equivalent to a $\mathbb{Z}$ linear relation between CM types of K . (Given r CM types of K , one has to choose E such that characters χ_i with values in E exist, corresponding to the types.) As mentioned before - III, 3.0 - nontrivial such relations exist if and only if K contains a CM field of degree at least 4.

1.1.1 In the situation of 1.1.0, let us assume, without loss of generality, that E is a CM field, and let us fix embeddings $K \overset{1}{\hookrightarrow} \mathbb{C}$, $E \overset{1}{\hookrightarrow} \mathbb{C}$ which allow us to consider K and E as subfields of $\mathbb{C}$.

There exist abelian varieties A_i with complex multiplication by E defined over K such that $H_1(A_i) = M(\chi_i)$ - see I, 4.1.1. If $n_i(\sigma,\tau) = -1$, there exists a holomorphic differential form $\omega^{(i)}_{\sigma,\tau}$ on A^σ_i , defined over $K^\sigma \cdot E^\tau \subset \mathbb{C}$, such that $e^*(\omega^{(i)}_{\sigma,\tau}) = e^\tau \cdot \omega^{(i)}_{\sigma,\tau}$, for all $e \in E$, and

$$(1.1.2) \qquad p(\chi_i;\sigma,\tau) = \int_{\gamma^{(i)}_\sigma} \omega^{(i)}_{\sigma,\tau}$$

(up to the usual indeterminacy), for any E basis $\gamma^{(i)}_\sigma$ of $H^\sigma_1(A_i)$ - cf. II, 1.5.1.

Putting $\sigma = 1$ in 1.1.2 shows that 1.0.1 implies Shimura's basic period relations, as stated, e. g., in theorems 1.2 and 1.3 of [Shi P]. Note, however, that the passage to antiholomorphic periods is normalized differently in Shimura's paper; he simply inverts the corresponding holomorphic period, whereas we are obliged to also multiply by $2\pi i$ - see II, 1.5.4.

The same translation also establishes propositions 1.4, 1.5, and 1.6 of [Shi P]. - Cf. [DP], 8.18 and [DB] for motivic interpretation and proof of all these relations in $\mathbb{C}^*/\overline{\mathbb{Q}}^*$.

Instead of explicitly stating these results up to $\overline{\mathbb{Q}}^*$, let us discuss finer relations provided by our formalism of the $p(\chi_i)$.

1.2 Shimura's refinement

1.2.0 As before, let $K \overset{1}{\hookrightarrow} \mathbb{C}$, $E \overset{1}{\hookrightarrow} \mathbb{C}$ be embedded algebraic number fields; E a CM field. Let χ be an algebraic Hecke character of K with values in E , and $n(\sigma,\tau)$ its Hodge exponents. Call $K_o \subset K$ the fixed field of

$$\{s \in G(\bar{\mathbb{Q}}/\mathbb{Q}) \mid n(s1,\tau) = n(1,\tau) \ , \text{ for all } \tau \in \mathrm{Hom}(E,\mathbb{C})\} \ .$$

Likewise, let $E_o \subset E$ be the fixed field of

$$\{s \in G(\bar{\mathbb{Q}}/\mathbb{Q}) \mid n(\sigma,s1) = n(\sigma,1) \ , \text{ for all } \sigma \in \mathrm{Hom}(K,\mathbb{C})\} \ .$$

K_o and E_o are each either $\mathbb{Q}$ or a CM field. From their definition, it follows that n descends to a function

$$n_o : \mathrm{Hom}(K_o,\mathbb{C}) \times \mathrm{Hom}(E_o,\mathbb{C}) \to \mathbb{Z}$$

such that

$$(*) \qquad n(\sigma,\tau) = n_o(\sigma|_{K_o},\tau|_{E_o}) \ ,$$

for all $\sigma \in \mathrm{Hom}(K,\mathbb{C})$, $\tau \in \mathrm{Hom}(E,\mathbb{C})$.

The following constructions will only depend on the function n_o , or equivalently, on <u>the algebraic homomorphism</u>

$$t : R_{K_o/\mathbb{Q}}\, \mathbb{G}_m \to R_{E_o/\mathbb{Q}}\, \mathbb{G}_m$$

defined by the $n_o(\sigma_o,\tau_o)$'s - see ⓪, § 2(c), and ⓪, § 4 . For all finite extensions $K_o \subset L$ and $E_o \overset{i}{\hookrightarrow} F$, the extended algebraic homomorphism

$$i \circ t \circ N_{L/K_o} : R_{L/\mathbb{Q}}\, \mathbb{G}_m \to R_{F/\mathbb{Q}}\, \mathbb{G}_m$$

is given by the function n on $\mathrm{Hom}(L,\mathbb{C}) \times \mathrm{Hom}(F,\mathbb{C})$ defined by equation (*). Thus, for $L = K$, $F = E$, we find the infinity type of the Hecke character χ . Writing w its weight, we have, for c = complex conjugation, that

$$n_o(\sigma_o,\tau_o) + n_o(\sigma_o,c\tau_o) = w ,$$

independent of (σ_o,τ_o) .

<u>1.2.1</u> There exists a finite extension $E_o \overset{i}{\hookrightarrow} F$, F a CM field, and a Hecke character χ_o of K_o with values in F and infinity type $i \circ t$. - Furthermore, there exists a finite abelian extension L of K_o such that $\tilde{\chi} = \chi_o \circ N_{L/K_o}$ takes values in E_o .

<u>1.2.2</u> Taken modulo $[(K_o^{\tilde{\sigma}})^{ab} \cdot E^{\tau_o}]^*$, the period $p(\tilde{\chi};\tilde{\sigma},\tau_o)$ - for $\tilde{\sigma} \in Hom(L,\mathbb{C})$, $\tau_o \in Hom(E_o,\mathbb{C})$ - depends only on $t,\sigma_o := \tilde{\sigma}|_{K_o}$, and τ_o . It will therefore be written

$$p(t;\sigma_o,\tau_o) \in \mathbb{C}^*/[(K_o^{\sigma_o})^{ab} \cdot E^{\tau_o}]^* .$$

If χ_i , for $i=1,\ldots,r$, are as in 1.1.0, and such that

$$(1.2.3) \qquad \mu \cdot \chi = \prod_{i=1}^{r} \chi_i ,$$

for some μ of finite order, then

$$(1.2.4) \qquad p(t;1,\tau_o) \underset{\overline{\mathbb{Q}}^*}{\sim} \prod_{i=1}^{r} p(\chi_i;1,\tau) ,$$

for any extension τ of τ_o to the common field of values E of χ and the χ_i . - Again, all that matters here, are the infinity types of the χ_i ...

Relation 1.2.4 almost establishes conjecture 1.7 in [Shi P]. The only difference is that Shimura wants the period which we denote by $p(t;1,\tau_o)$ to be well defined up to $[K_o^{ab} \cdot E_o^{+\tau_o}]^*$, where E_o^+ is the maximal totally real subfield of E_o . To achieve this, we simply repeat remark 8.22 of [DP] in our context:

First, note the naturality of the formation of $p(t)$, which is easily proved on the level of Hecke characters, by transport of structure - cf. [DP], 8.18.4:

<u>1.2.5</u> If $\alpha : K_0' \xrightarrow{\cong} K_0$ and $\beta : E_0 \xrightarrow{\cong} E_0'$ are isomorphisms of fields, then

$$p(\beta t\alpha;\sigma_0\alpha,\tau_0\beta^{-1}) = p(t;\sigma_0,\tau_0) \ .$$

From this, II, 1.6.6, and the fact that complex conjugation c induces a well defined automorphism of K_0 as well as E_0, we obtain the equations

$$(1.2.6) \qquad \overline{p(t;\sigma_0,\tau_0)} = p(t;c\sigma_0,c\tau_0) = p(t;\sigma_0 c,\tau_0 c)$$

$$= p(ctc;\sigma_0,\tau_0) = p(t;\sigma_0,\tau_0) \ .$$

(Note that $(K_0^{\sigma_0})^{ab}$ is stable under c, even though c does not in general commute with other automorphisms of $(K_0^{\sigma_0})^{ab}$!)

Thus, by Hilbert 90, <u>the periods</u> $p(t;\sigma_0,\tau_0)$ <u>are represented by real numbers; they are well determined up to a factor in</u> $[(K_0^{\sigma_0})^{ab}E_0^{+\tau_0}]^*$.

<u>1.2.7 Remark.</u> D. Blasius has informed me that he has not only found the above results independently, by the motivic formalism; but that he has also managed to prove conjecture 1.7 of [Shi P] adapting Shimura's proof - as in [Shi P], section 5 - , thus improving upon his 1981 Princeton thesis (unpublished) in which a partial result was obtained.

Let us now go on to examine a few standard properties of the $p(t;\sigma_0,\tau_0)$ - cf. [DP], 8.18 for the relations up to $\overline{\mathbb{Q}}^*$.

<u>1.2.8</u> For finite extensions $K_0 \subset K_0'$, $E_0 \overset{i}{\hookrightarrow} E_0'$, E_0 a CM field, put

$$p(i \circ t \circ N_{K_0'/K_0};\sigma',\tau') = p(t;\sigma'|_{K_0},\tau'|_{E_0}) \ ,$$

but consider the left hand side as being well determined up to $[(K_0'^{\sigma'})^{ab}(E_0'^{+})^{\tau'}]^*$. - Cf. II, 1.8.6/8.

<u>1.2.9</u> If u and u' are two algebraic homomorphisms $K^* \to E^*$, both satisfying $n^{(')}(\sigma,\tau) + n^{(')}(\sigma,c\tau) = cst$, independent of (σ,τ), then - in the sense of 1.2.8 - ,

$$p(u;\sigma,\tau)\cdot p(u';\sigma,\tau) = p(u\cdot u';\sigma,\tau) \ .$$

1.2.10 From 1.2.9 and II, 1.8.4, we get:

$$p(t;\sigma_o,\tau_o)\cdot p(t;\sigma_o,c\tau_o) = (2\pi i)^w \ ,$$

where

$$w = n_o(\sigma_o,\tau_o) + n_o(r_o,c\tau_o).$$

1.2.11 "Reflex Principles"

"Reflex principles" - like theorem 2.3 of [Shi P]; theorem 1.2 of [Shi O]; proposition 8.20 of [DP] - are formal consequences of the general formalism of the periods concerned - i.e., in our case, of formulas 1.2.5, 1.2.8 - 1.2.10. They typically serve to trace the periods through arguments in which the rôles of the fields K_o and E_o above are interchanged. See, for instance, the use that Deligne makes of [DP], 8.20; cf. [HS], p. 35.

Theorem 1.2 in [Shi O] has the advantage of being particularly simple and general. But it seems to require a formalism of periods that are essentially only determined up to a factor in $\overline{\mathbb{Q}}^*$. - So, we content ourselves with a refined version of [DP], 8.20.

Let $t^* : R_{E_o/\mathbb{Q}}\,\mathbb{G}_m \to R_{K_o/\mathbb{Q}}\,\mathbb{G}_m$ be the algebraic homomorphism defined by the invariants $n_o^*(\tau_o,\sigma_o) = n_o(\sigma_o,\tau_o)$. Define $p(t^*;\tau_o,\sigma_o)$ as in 1.2.1/2, via $p(\widetilde{\chi}^*;\widetilde{\tau},\sigma_o)$, for a suitable Hecke character $\widetilde{\chi}^*$ of some finite abelian extension F^* of E_o with values in K_o. Assume, as in II, 1.8.1, that none of the $n_o(\sigma_o,\tau_o)$'s equals $\frac{1}{2}w = \frac{1}{2}(n_o(\sigma_o,\tau_o) + n_o(\sigma_o,c\tau_o))$. Then we have the following identity of classes in $\mathbb{C}^*/(M^{ab})^*$, where M is the smallest Galois extension of $\mathbb{Q}$ contained in $\mathbb{C}$ which contains both K_o and E_o.

$$(1.2.12) \qquad \prod_{\substack{\sigma_o \\ n_o(\sigma_o,1)<\frac{w}{2}}} p(t^*;1,\sigma_o) = \prod_{\substack{\sigma_o \\ n_o(\sigma_o,1)<\frac{w}{2}}} p(t;\sigma_o,1) \ .$$

The proof is straightforward - cf. [DP], 8.20. - Further refinements of 1.2.12 may be treated using the construction of Blasius mentioned in II, 2.2 ...

1.2.13 For examples of these period relations, the reader should consult section 2 of [Shi P] . - As will be indicated in § 2, the precise relations of the form (1.0.1) are liable, in principle, to yield much more information in concrete circumstances (in particular, if μ can be computed) than the coarser but more flexible periods $p(t;\sigma,\tau_0)$ with their increased indeterminacy.

2. Abelian integrals and the gamma function

2.0 In this section we consider a finite imaginary abelian extension K of $\mathbb{Q}$, with Galois group G. We fix a privileged embedding $K \overset{1}{\hookrightarrow} \mathbb{C}$.

2.1 Let χ be any algebraic Hecke character of K (with values in some number field E). Write its infinity type as

$$(2.1.0) \qquad t = \sum_{\sigma \in G} n_\sigma \sigma \in \mathbb{Z}[G] .$$

By ●, 8.4.2, there exists for χ,

- a smallest positive (or: nonzero of smallest absolute value) integer h, such that there is
- an element $\underline{a} \in \mathbb{B}_K^0$, - see ●, 8.2.1 -
- and a character of finite order μ of K,

satisfying,

$$(2.1.1) \qquad \mu \cdot \chi^h = J_K(\underline{a}) .$$

Thus, $\mu \cdot \chi^h$ takes values in K, and via I, 5.1 we can translate 2.1.1 into the period relation

$$(2.1.2) \qquad p(\mu\chi^h;1) = (\Gamma(\sigma c\underline{a})^{-1})_{\sigma \in G} \in (K \otimes \mathbb{C})^*/K^* ,$$

using II, 4.2.3 and II, 4.4.2.

2.1.3 The left hand side of 2.1.2 is easily expressed in terms of $p(\mu)$ and $p(\chi)$; so any information one has about μ can, in principle, serve to relate $p(\chi)$ to values of the gamma function.

But already the very explicit case discussed in chapter III shows that μ will usually not be easy to determine from χ, h and $\underline{a}$.

2.1.4 Therefore, let us now discard finite order characters in 2.1.1. Then $\underline{a}$ is determined by χ and h up to addition of an element $\underline{b} \in \mathbb{B}_K^0$ such that $J_K(\underline{b})$ is of finite order. In other words, $\underline{a}$ is such that, for all $\sigma \in G$, one has

$$(2.1.5) \qquad \frac{1}{h}\langle \sigma^{-1}\underline{a}\rangle = n_\sigma .$$

By II, 4.4.5, this determines $(\Gamma(\sigma c\underline{a}))_{\sigma \in G}$ up to a factor in $(K^{ab})^* \subset \overline{\mathbb{Q}}^* \subset \mathbb{C}^*$.

Using 1.2.6 and the notation of 1.2.8 above, it follows for all infinity types t as in 2.1.0 and all $\underline{a} \in \mathbb{B}_K^0$ with 2.1.5 that, for all $\sigma \in G$, we have the period relation in $\mathbb{R}^*/(K^{ab} \cap \mathbb{R})^*$,

$$(2.1.6) \qquad p(t;1,\sigma)^h = \Gamma(\sigma c\underline{a})^{-1} .$$

2.1.7 Considering both sides of 2.1.6 as representing classes in $\mathbb{C}^*/\overline{\mathbb{Q}}^*$ this relation verifies Gross' period conjecture - see [Gr 2] § 4 - for all motives in $\mathcal{M}_{\overline{\mathbb{Q}}}^{av}(K)$ of rank 1 over K. In fact, all these motives are determined by their Hodge realization (say, at $id_{\overline{\mathbb{Q}}} : \overline{\mathbb{Q}} \hookrightarrow \mathbb{C}$); but they all come from motives of the form $M(\chi)$ in $\mathcal{M}_K^{av}(E)$, for some χ as above - see I, 5.3; I, 6.1.4, and I, 6.1.6. - This remark covers (and therefore indicates possible refinements of) the first two examples discussed by Gross in [Gr 2] p. 206/7; as for example 3 (p. 207/8), a motivic version of it will be established in chapter V below.

2.1.8 Suppose $\mathcal{C}$ is a class of (smooth projective) $\overline{\mathbb{Q}}$-algebraic varieties for each of which one can show that every Hodge cycle on it is an absolute Hodge cycle. For $L \subset \overline{\mathbb{Q}}$, let $\mathcal{M}[\mathcal{C}]_L$ be the smallest Tannakian subcategory of $\mathcal{M}_L$ which contains $C\mathcal{M}_L$ as well as all motives of the form $h(X_L)$, where X_L is a variety over L which becomes isomorphic, over $\overline{\mathbb{Q}}$, to a variety in $\mathcal{C}$. Then Gross' period conjecture holds for the Hodge realizations of all motives in

$\mathcal{M}[\mathcal{C}]_{\overline{\mathbb{Q}}}(K)$ of rank 1 over K . Moreover, the results of this section would then extend to all motives in $\mathcal{M}[\mathcal{C}]_K(E)$ attached to Hecke characters χ of K , in the sense of I, 3.3.- This follows from our proof of I, 5.1; from I, 5.3 and I § 6.

<u>2.1.9</u> We saw in 2.1.5 that working modulo finite order characters in 2.1.1 allows to express $\underline{a}$ in terms of χ and h . As for h , let us simply remark that it may always be replaced with the index $[A_K : St_K]$ of **0**, 8.4. We refer to [Sin] for a number of results on this index.

<u>2.2</u> Looking at things the other way around, let now $\underline{a} \in \mathbb{B}_K^0$ be given. Then there exist

- a positive integer r
- algebraic Hecke characters $\chi_1,\ldots,\chi_r$ of K like in 1.1.0 above
- integers $n_1,\ldots,n_r$
- a character of finite order μ of K

such that

$$(2.2.1) \qquad \mu \cdot \prod_{i=1}^{r} \chi_i^{n_i} = J_K(\underline{a}) .$$

Let E be a finite extension of K which is a common field of values for μ and all the χ_i . Write j the inclusion $(K \otimes \mathbb{C})^* \hookrightarrow (E \otimes \mathbb{C})^*$. Then 2.2.1 translates into the period relation

$$(2.2.2) \qquad p(\mu;1) \cdot \prod_{i=1}^{r} p(\chi_i;1)^{n_i} = j((\Gamma(\sigma c\underline{a})^{-1})_{\sigma \in G}) .$$

By section 1 above, this relation "does not depend" on the particular choice of χ_i's - in the sense that all such relations arising from $\underline{a}$ can be seen to be equivalent without recourse to the right hand sides . - Conversely, all Γ-relations arising from comparing the left hand sides of two instances of 2.2.2, for two different $\underline{a},\underline{a}' \in \mathbb{B}_K^0$, via 1.0.1 already follow from II, 4.4.5, and are therefore, if taken modulo $\overline{\mathbb{Q}}^*$, compatible with Rohrlich's conjecture - see II, 4.0 - , by virtue of the remark we made following theorem

II, 4.4.5. Shimura's feelings about this, expressed at the end of [Shi P] § 4, are therefore proven to have been correct.

We leave it to the reader to write weaker versions of 2.2.2, neglecting finite order characters in 2.2.1.

2.3 Biquadratix

Let $K \hookrightarrow \mathbb{C}$ be an abelian imaginary field of degree four over $\mathbb{Q}$.

2.3.1 If K is cyclic, then there is a simple abelian variety A with complex multiplication by K , defined, say, over K^{ab} , and all such simple abelian surfaces are isogenous to some conjugate of A . 2.2.2 relates certain products of Γ values to periods of the A^{σ}'s. E. g., when $K = \mathbb{Q}(\mu_5)$, one finds precisely the well known expressions of the periods of the Jacobian of $X^5+Y^5 = 1$ in terms of Γ - cf. II, 4.1, or Rohrlich's appendix to [Gr 2].

2.3.2 If K is not cyclic, call K^+ its real quadratic subfield, and F_1 , F_2 the two distinct imaginary quadratic subfields of K . Every abelian variety A with complex multiplication by K is isogenous to the product of two CM elliptic curves - cf. [Sch A] for the exceptional rôle that such K play among all CM fields. All Hecke characters χ of K like in 1.1.0 can be written as

$$(2.3.3) \qquad \chi = \mu(\varphi_i \circ N_{K/F_i}) \quad ,$$

for some $i \in \{1,2\}$, and some Hecke character φ_i of F_i of infinity type σ_i , for some embedding $\sigma_i : F_i \hookrightarrow \mathbb{C}$; and μ some character of finite order of K . The only nontrivial relations of the form 1.0.0 over K are consequences of the relations $\varphi_i\overline{\varphi}_i = \mathbb{N}^{-1}$ over F_i . Thus, the corresponding period relations reproduce Legendre's relation on elliptic curves with complex multiplication by F_i - cf. II, 1.8.5. Similarly, it is easy to check that 2.2.2 does not produce any new relations between the elliptic integrals with CM by F_1 or F_2 , and Γ values, beyond what follows from II § 4 and III, 1.4 - cf. III § 3 .

CHAPTER FIVE:

Motives of CM Modular Forms

Let K be a CM field with maximal totally real subfield K_o. Given a Hecke character χ of K with corresponding theta series f (a Hilbert modular new-form relative to K_o), there should be a motive $M(f)$ for f whose periods could be computed in terms of special values of $L^*(\chi,s)$. If one could not only construct the motive $M(f)$ in $\mathcal{M}_{\mathbb{Q}}(E_o)$ - with E_o the field generated by the Fourier coefficients of f - but also show that, since f comes from χ, $M(f)$ lies in $\mathcal{M}_{\mathbb{Q}}^{av}$, and in fact in $\mathcal{CM}_{\mathbb{Q}}$, then theorem I, 5.1 would allow to compare $M(f)$ and $M(\chi)$, and thereby yield II, 2.1 for χ (or closely related characters), provided certain non vanishing results are available, about the special values of $L(\chi,s)$ mentioned before.

This hypothetical "modular proof" of II, 2.1 seems a long way off at the moment - cf. Oda's work [Od 1], [Od 2]. However, it provides the romantic background for what we actually prove in this chapter: First of all, we consider only the case that K is imaginary quadratic. In this case, recent observations of U. Jannsen's, in connection with his more general theory of mixed motives, easily give us the actual motive $M(f)$ whose realizations were already described in [DP] § 7 - this is discussed in § 1 below. Then, after introducing the theta series for Hecke characters of K we do prove, in § 2, that $M(f)$ lies in $\mathcal{CM}_{\mathbb{Q}}$. But in order to do so, we have to use II, 2.1 for Hecke characters of K - in this case the theorem was first proved in [GS], [GS'].

1. Motives for modular forms

Let $k \geq 0$ and $N \geq 1$ be integers. Let $f(z) = \sum_{n \geq 1} a_n q^n$ ($q = e^{2\pi i z}$) be a newform on $\Gamma_o(N)$ of weight $k+2$ with character ε, which is an eigenfunction for the Hecke operators T_p, p prime, $p \nmid N$:

$$T_p f = a_p f \; ; \quad a_1 = 1 .$$

Put $E_o = \mathbb{Q}(a_n \mid (n,N) = 1) \hookrightarrow \mathbb{C}$. It is known that E_o is a number field of finite degree over $\mathbb{Q}$.

1.1 Theorem [Eichler-Shimura-Deligne-Jannsen]

There exists a motive $M(f)$ *in* $\mathcal{M}_{\mathbb{Q}}(E_o)$ *of rank two over* E *such that*

$$L_N^*(M(f),s) := \Big(\prod_{p\nmid N} \det_{E_o}(1-F_p\cdot p^{-s}|H_\ell(M(f)))^{-\tau}\Big)_\tau = \Big(\sum_{\substack{n=1\\(n,N)=1}} a_n^\tau n^{-s}\Big)_\tau ,$$

where $\tau \in \mathrm{Hom}(E_o,\mathbb{C})$, $\ell \neq p$ *prime*, F_p *is a geometric Frobenius element at* p , *and* $\mathrm{Re}\, s \gg 0$.

The proof of this theorem is indicated in [Ja], Cor. 1.4, building upon [DR] and [DP] § 7. Let us sketch very briefly how one can show that the realizations written in [DP], 7.6 actually are realizations of a motive in $\mathcal{M}_{\mathbb{Q}}$, by using a somewhat different argument - which, however, was also suggested to me by Jannsen.

1.1.1 Write $A_o = Y_1(N)$, and $\overline{A}_o = X_1(N)$ the modular curves without, and with cusps. Suppose $N \geq 3$ and denote by $\pi_1 : A_1 \to A_o$ the universal elliptic curve. Put

$$A_k = \underbrace{A_1 \times_{A_o} \cdots \times_{A_o} A_1}_{k \text{ factors}}$$

if $k \geq 2$. Choose a smooth compactification $\overline{A}_k$ of A_k - cf., for instance [DR], 5.5. Let Z be a desingularization of $\overline{A}_k \setminus A_k$. Then, by [DH III], Cor. 8.2.8, one has in the diagram

$$\begin{array}{ccccc} H_c^{k+1}(A_k,\mathbb{Q}) & \xrightarrow{\beta} & H^{k+1}(\overline{A}_k,\mathbb{Q}) & \xrightarrow{\alpha} & H^{k+1}(\overline{A}_k\setminus A_k,\mathbb{Q}) \\ & & & \searrow^{\tilde{\alpha}} & \downarrow \\ & & & & H^{k+1}(Z,\mathbb{Q}) \end{array}$$

that

$$H^{k+1}_{!}(A_k,\mathbb{Q}) := \mathrm{Im}(\beta) = \ker(\alpha) = \ker(\tilde{\alpha}) .$$

As $H^{k+1}(\bar{A}_k)$ and $H^{k+1}(Z)$ are honest regard motives, and $\tilde{\alpha}$ comes from an absolute Hodge cycle $\tilde{\alpha} : H^{k+1}(\bar{A}_k) \to H^{k+1}(Z)$, its kernel, too, defines a motive in $\mathcal{M}_{\mathbb{Q}}$, inside of which one now continues to cut out the desired submotive:

1.1.2 By Lieberman 's trick - see [DR], 5.3 -, preferably modified by taking the part where $[m_1] \times \ldots \times [m_k]$ act as $m_1 \cdot \ldots \cdot m_k$, for sufficiently many collections of integers $(m_1, \ldots, m_k)$, one obtains a motive with realizations

$$H^1_!(A_o,(R^1\pi_{k,*}\mathbb{Q})^{\otimes k}) .$$

1.1.3 Next, take invariants under the action of the symmetric group S_k , and finally pass to the submotive of

$$H^1_!(A_o,\mathrm{Sym}^k(R^1\pi_{k,*}\mathbb{Q})) = H^1(\bar{A}_o,j_*\mathrm{Sym}^k(R^1\pi_{k,*}\mathbb{Q}))$$

annihilated by the kernel of the homomorphism of the Hecke algebra $\mathbb{T}$,

$$\begin{array}{ccc} \mathbb{T} & \to & E_o \\ T_p & \to & a_p . \end{array}$$

This produces the realizations described in [DP], 7.6.

1.1.4 In case N was 1 or 2, or if we want to construct motives for modular forms on more general congruence subgroups, one has to close the construction of M(f) by passing to the invariants under a finite subgroup.

It is fairly clear that, at every stage, we have only applied absolute Hodge cycles in cutting out the next smaller motive. But we leave the details to the reader.

The following problem seems to be unsolved.

1.2 Problem. Show that for "generic" f of weight $k+2 \geq 3$ - and in particular for $\Delta(z) = q\prod_1^{\infty}(1-q^n)^{24}$? - M(f) does not lie in $\mathcal{M}_{\mathbb{Q}}^{av}(E_o)$.

1.2.1 For $k = 0$, that is, if f has weight two, M(f) "is" **essentially the abelian variety attached to f by Shimura - see, e.g., [Shim], Thm. 7.14. So, $M(f) \in \mathcal{M}_{\mathbb{Q}}^{av}$ for $k = 0$.**

2. CM modular forms

2.0 Let K be an imaginary quadratic field, embedded $K \xrightarrow{1} \mathbb{C}$ in a fixed way, and write $-D$ the discriminant of K . Let χ be an algebraic Hecke character of K , of conductor $\mathfrak{f}$, with infinity type $w\cdot 1$, for some $w \geq 1$. Denote by $E \supset K$ the number field generated by the values of χ . Write the theta series f attached to χ , and an embedding $\tau : E \hookrightarrow \mathbb{C}$:

$$f^\tau(z) = \sum_{(\mathfrak{a},\mathfrak{f})=1} \chi^\tau(\mathfrak{a})q^{\mathbb{N}\mathfrak{a}} = \sum_{n\geq 1} a_n^\tau q^n ,$$

where $\mathfrak{a}$ runs over all integral ideals of K prime to $\mathfrak{f}$, and $a_n = \sum_{\mathbb{N}\mathfrak{a}=n} \chi(\mathfrak{a})$ - thus, $a_1 = 1$. - By Hecke, [He], n° 23, 27, $f^\tau(z)$ is, for each τ , a newform of weight $w+1$ (i.e., $k = w-1$, in the notation of § 1) on $\Gamma_o(N)$, with $N = D\cdot\mathbb{N}\mathfrak{f}$, and character ε given on prime numbers $p \nmid N$ by

$$\varepsilon(p) = \left(\frac{-D}{p}\right)\frac{\chi(p\cdot\mathcal{O}_K)}{p^w} .$$

2.1 As in 1, call E_o the field $\mathbb{Q}(a_n|(n,N)=1)$, and write $\mathbb{Q}(\varepsilon)$ the field generated by the values of ε .

2.1.1 Lemma (i) $E = K\cdot E_o$

(ii) $E_o \supset \mathbb{Q}(\varepsilon)$.

The proof is left to the reader.

Three cases occur, as far as the constellation $E_0 \subset E$ is concerned:

2.1.2 E_0 is totally real.

This is the same as saying that E, which is naturally a CM field, has E_0 as its maximal totally real subfield. Also, like in 2.1.1 (i), one sees that 2.1.2 occurs if and only if χ is equivariant with respect to complex conjugation, i. e.,

$$(2.1.2)' \qquad \chi(\bar{\mathfrak{a}}) = \overline{\chi(\mathfrak{a})} ,$$

for all $(\mathfrak{a}, \mathbb{N}\mathfrak{f}) = 1$.

Note also that $\epsilon = 1$ implies 2.1.2, and that 2.1.2 in turn yields $\epsilon^2 = 1$ (but not necessarily $\epsilon = 1$, as is shown by the example of χ with (2.1.2)' of conductor (1).)

2.1.3 $E = E_0$.

By 2.1.1 (i), if E_0 is not totally real, E_0 has to be a CM field. But NOT 2.1.2 does not imply 2.1.3 because it may also be that:

2.1.4 $E \neq E_0$ and E_0 is not totally real.

As an example (pointed out to me by J. Tilouine), take χ satisfying (2.1.2)', and ν a Dirichlet character of $\mathbb{Q}$ such that $K \not\subset \mathbb{Q}(\nu(\mathbb{N}\mathfrak{a}) \mid \mathfrak{a}$ ideal of $K)$. Then, for the twist $\chi\cdot(\nu \circ N_{K/\mathbb{Q}})$, one is in case 2.1.4.

2.2 Proposition Let f be as in 2.0 and E_0 as defined in 2.1. Let $M(f)$ be the motive attached to f by 1.1. Then there is a natural embedding

$$K \otimes_{\mathbb{Q}} E_0 \hookrightarrow \mathrm{End}_{/K}(M(f) \times K) ,$$

inducing on $1 \otimes E_0$ the coefficient structure of $M(f) \in \mathcal{M}_{\mathbb{Q}}(E_0)$, and such that, for every idempotent e of $K \otimes E_0$ with $e(K \otimes E_0) \cong E$, the direct factor $e(M(f) \times K)$ of $M(f) \times K$ is a motive either for χ or for the complex conjugate $\bar{\chi}$, in the sense of I, 3.3.

<u>Proof.</u> We shall essentially generalize Shimura's proof of theorem 1 in [Shi E]. (For the broader perspective of this method cf. [Shi F], [Ri 1], [Mom], and [Ri 2].)

First assume that the conductor $\mathfrak{f}$ of χ satisfies:

(2.2.0) $$D \mid \mathfrak{f} \quad \text{and} \quad \overline{\mathfrak{f}} = \mathfrak{f} .$$

In analogy to the definition of the Hecke operators, the double class $\Gamma_1(N)\,\delta\,\Gamma_1(N)$, with

$$\delta = \begin{pmatrix} 1 & \frac{1}{D} \\ 0 & 1 \end{pmatrix} \in SL_2(\mathbb{Q}) ,$$

induces an algebraic correspondence on $X_1(N) \times X_1(N)$. This can be lifted to A_k and closed up in $\overline{A}_k$, and actually induces an (absolute-Hodge-) endomorphism Δ of the direct sum $\underset{\lambda}{\oplus} M(f_\lambda)$, with λ running through all algebraic Hecke characters of K defined modulo $\mathfrak{f}$, defined over $\mathbb{C}$. To see this, note that for any such λ, the corresponding theta series

$$f_\lambda(z) = \sum_{(\mathfrak{a},\mathfrak{f}) = 1} \lambda(\mathfrak{a}) \cdot q^{\mathbb{N}\mathfrak{a}}$$

can be written

$$f_\lambda(z) = \sum \lambda(\mathfrak{b})^{-1} F_\mathfrak{b}(z) ,$$

where $\mathfrak{b}$ varies over a fixed system of integral ideals of K representing the ray classes of ideals prime to $\mathfrak{f}$ modulo principal ideals (α) generated by elements $\alpha \equiv 1 (\mathrm{mod}\, \mathfrak{f})$ in K^* , and where

$$F_\mathfrak{b}(z) = \sum_{\substack{(\alpha) \\ \alpha \in \mathfrak{b}}} \alpha^w \cdot q^{\mathbb{N}(\alpha)/\mathbb{N}(\mathfrak{b})} .$$

Thus, by the Shimura isomorphism, $H_{DR}(\underset{\lambda}{\oplus} M(f_\lambda)/\mathbb{C})$ is generated by the $F_\mathfrak{b}$'s (and their antiholomorphic counterparts); but on them Δ acts via

$$(2.2.1)\qquad \Delta_{DR}(F_b)(z) = \sum_{j=1}^{r} F_b\big|_{w+1}(\delta\gamma_j) = r\cdot(e^{2\pi i/D})^{\frac{\mathbb{N}\alpha}{\mathbb{N}b}}\cdot F_b\ ,$$

if $\Gamma_1(N)\delta\Gamma_1(N) = \bigcup_{j=1}^{r} \Gamma_1(N)\delta\gamma_j$. Here, as $D|\mathfrak{f}$, by 2.2.0,

$\mathbb{N}\alpha/\mathbb{N}b$ is independent of the choice of $\alpha \in b$, $\alpha \equiv 1(\mathrm{mod}\,\mathfrak{f})$.

2.2.2 Note that, for the case 2.2.0, it would have been sufficient to let λ above vary over the characters of precise conductor $\mathfrak{f}$. But, for future reference, we throw in all λ defined modulo $\mathfrak{f}$ - understanding that, for them, the motive denoted $M(f_\lambda)$ is, for once, not the motive $M(f_\lambda)$ constructed in § 1, but rather the motive obtained by the procedure sketched in 1.1.1-3 for f_λ , considered as an eigenform on $\Gamma_1(N)$ - where it is not a newform. The proper motive for f_λ in the sense of § 1 is found inside the current $M(f_\lambda)$ as the invariants under the finite group which is the quotient of $\Gamma_1(N)$ by the group $\Gamma_1(M)$ on which f_λ "is new" - cf. 1.1.4.

2.2.3 From 2.2.1, we see that we can embed $\mathbb{Q}(e^{2\pi i/D})$ into $\mathrm{End}_{/\mathbb{C}}(\bigoplus_\lambda M(f_\lambda))$ by sending $e^{2\pi i/D}$ to $r^{-1}\cdot\Delta$. The imaginary quadratic field $K \subset \mathbb{Q}(e^{2\pi i/D})$ then induces endomorphisms which stabilize $M(f) \subset \oplus M(f_\lambda)$, as is checked again on the de Rham realization, using 2.2.1. **Next, it is easy to see that this embedding**

$$K \hookrightarrow \mathrm{End}\, M(f)$$

is defined over K .

Moreover, the commutator of the matrices $\begin{pmatrix}1 & 1/D\\ 0 & 1\end{pmatrix}$ and $\begin{pmatrix}1 & 0\\ 0 & 1/p\end{pmatrix}$ lies in $\Gamma_1(N)$ if the prime p splits completely in $\mathbb{Q}(\mu_D)$. As T_p is 0 on $M(f)$ for all p that stay prime in K, we see that the action of K on $M(f)\times K$ we defined commutes with the action of E_o.

2.2.4 Thus we know - in case 2.2.0 - that $M(f)\times K$ is of rank 1 over $K\otimes E_o$. Since the actions of K and E_o come from algebraic correspondences , we see that the one dimensional Galois representations of $e(M(f)\times K)$ form a strictly compatible system of E-rational λ-adic representations - where λ now denotes finite

places of E . In view of the L-function of $M(f)$ over $\mathbb{Q}$, the Hecke character defined by $M(f) \times K$ - see I, 1.4 - has to be either χ or $\overline{\chi}$.

2.2.5 If $\mathfrak{f}$ is arbitrary, we consider χ as a character defined modulo $D \cdot \mathfrak{f} \cdot \overline{\mathfrak{f}}$. It occurs then as one of the imprimitive characters λ of the above argument, and 2.2.3 and 2.2.4 show that "$M(f)$" = $M(f_\lambda)$ - in the sense 2.2.2 - has the required $K \otimes E_0$ structure. Now, this K-action clearly commutes with the finite group G such that $M(f_\lambda)^G$ is the proper motive $M(f)$ of the newform f .

q.e.d.

2.3.0 Let ψ be any Hecke character of K , with values in some CM field E' , of conductor $\mathfrak{f}'$, with infinity type 1 , such that $L^*(\psi,1) \in (E' \otimes \mathbb{C})^*$. The existence of such ψ is most easily deduced from the fact that the modular symbols generate the first homology of the modular curves - see [Shi M], theorem 2. Using this argument, we have already passed to the newform $g(z) = \sum \psi(\mathfrak{a}) q^{\mathbb{N}\mathfrak{a}} = \sum b_n q^n$ on $\Gamma_0(N')$, $N' = D \cdot \mathbb{N}\mathfrak{f}'$, associated to ψ as f is to χ in 2.0. Write E_0' the field generated by the Fourier coefficients of g .

2.3.1 Let μ be the finite order character of K such that

$$\chi = \mu \cdot \psi^w ,$$

and $\mathbb{Q}(\mu)$ its field of values. By 2.2, the motive $M(g) \times K$ has a natural $K \otimes E_0'$ action. Calling $\widetilde{E}$ the composite $E_0 \cdot E_0' \cdot \mathbb{Q}(\mu)$ - in some fixed algebraic closure of $\mathbb{Q}$ - we get the motives with coefficients in $K \otimes_{\mathbb{Q}} \widetilde{E}$ defined over K :

- $\widetilde{M}(\mu) = M(\mu) \otimes_{\mathbb{Q}(\mu)} (K \otimes \widetilde{E})$
- $\widetilde{M}(g) = (M(g) \times K) \otimes_{K \otimes E_0'} (K \otimes \widetilde{E})$
- $\widetilde{M}(f) = (M(f) \times K) \otimes_{K \otimes E_0} (K \otimes \widetilde{E})$

They are all of rank 1 over $K \otimes \widetilde{E}$. - For the next theorem we suppose that the K actions on $M(f) \times K$ and $M(g) \times K$ have been normalized so that, for every idempotent $\widetilde{e}$ of $K \otimes \widetilde{E}$ - cf. 2.2 - , the factor $\widetilde{e}(\widetilde{M}(f))$ is a motive for χ (with values in $\widetilde{E} \cdot E$) if and only if $\widetilde{e}(\widetilde{M}(g))$ is a motive for ψ (with values in $\tilde{E}\cdot E$) .

2.4 Theorem. There is an isomorphism of motives with coefficients in $K\otimes\tilde{E}$, defined over K ,

$$\tilde{M}(\mu)\otimes_{K\otimes\tilde{E}}\tilde{M}(g)^{\otimes_{K\otimes\tilde{E}}w}\cong\tilde{M}(f) .$$

Proof. First, note that, if $w=1$, then $\tilde{M}(f)$ and $\tilde{M}(g)$ both lie in $\mathcal{M}_K^{av}$ - cf. 1.2.1 - , and I, 5.1 gives us the isomorphism of the theorem. - For $w\geq 2$, we shall construct this absolute Hodge correspondence via the relation between periods and L-values: By 1.1 and the construction of f and g , we have

$$L_N^*(\tilde{M}(f),s)=\left(\sum a_n^\tau n^{-s}\right)_{\tau:\tilde{E}\hookrightarrow\mathbb{C}}=L^*_{Df\bar{f}}(\chi,s)$$

$$L_{N'}^*(\tilde{M}(g),s)=\left(\sum b_n^\tau n^{-s}\right)_{\tau:\tilde{E}\hookrightarrow\mathbb{C}}=L^*_{Df'\bar{f}'}(\psi,s) ,$$

where, on the right, we have written the L-functions of 0 § 6 for the characters χ and ψ , considered as taking values in $\tilde{E}$, deleting the Euler factors above N , resp. N' . Such Euler factors, taken at critical integers s , lie in $\tilde{E}^*$, and will therefore be disregarded in the argument that follows.

Since $w\geq 2$, it is well-known that no component of $L^*(\chi,w)$ vanishes. Also, $L(\psi^\tau,1)\neq 0$, for all τ , by construction. Therefore, as $\chi=\mu\cdot\psi^w$, it follows from II, 2.1 - which, for K imaginary quadratic, was already proved in [GS] and [GS'] - , using II, 1.8.1/3 and II, 1.7.12 (iv), that

(2.4.0) $\quad L^*(\tilde{M}(\mu)\otimes\tilde{M}(g)^{\otimes w},w)=L^*(\tilde{M}(f),w)\in(\tilde{E}\otimes\mathbb{C})^*$,

up to a factor in $(\tilde{E}\otimes 1)^*$.

Now, as Deligne points out - [DP], 7.6 - the motive $M(f)$ is constructed in such a way that

$$L^*(\tilde{M}(f),w)=c^+(R_{K/\mathbb{Q}}\tilde{M}(f)(w))\in(\tilde{E}\otimes\mathbb{C})^*/\tilde{E}^* .$$

Similarly, the analogous relation for the left hand side of 2.4.0 follows from [DP], 7.2 - or from the observation that $\widetilde{M}(\mu) \otimes \widetilde{M}(g)^{\otimes w}$ is in $\mathcal{M}_K^{av}$; cf. the first sentence of this proof.

As K is imaginary quadratic, 2.4.0 then implies via II, 1.7.3-6 - note that M(f) is of Hodge type (w,0) + (0,w) so that II, 1.7 applies! - that

$$p(\widetilde{M}(\mu) \otimes \widetilde{M}(g)^{\otimes w}) = p(\widetilde{M}(f)) .$$

Thus we have shown that the two motives to be compared in 2.4 are <u>of rank</u> 1 <u>over</u> $K \otimes \widetilde{E}$, defined over K ; they are motives for the <u>same</u> algebraic Hecke character - to wit, $\mu \cdot \psi^w = \chi$ - , in the sense of I, 3.3, extended to our situation where the coefficient algebra may be a product of fields; and they have the <u>same</u> periods p . **Since we are in a rank 1 situation, this is sufficient to physically construct an absolute Hodge correspondence between them establishing their isomorphism: all it comes really down to is choosing bases - i. e., each time a non trivial element - for the various realizations of the two motives.**

q.e.d.

<u>2.4.1 Remark.</u> Richard Pink, in an unpublished note, has shown that the absolute Hodge correspondence we just constructed is actually an <u>algebraic cycle</u>, in the special case where $K = \mathbb{Q}(\sqrt{-4})$, ψ is the Hecke character of the elliptic curve $y^2 = 4x^3 - 4x$ - cf. I, 7.5 - , or, in other words, of $X_0(32)$, and $\chi = \psi^2$.

<u>2.4.2 Corollary .</u> M(f) <u>"lies in"</u> $C\mathcal{M}_{\mathbb{Q}}(E_0)$, <u>in the sense that it is isomorphic in</u> $\mathcal{M}_{\mathbb{Q}}(E_0)$ <u>to an object of</u> $C\mathcal{M}_{\mathbb{Q}}(E_0)$; <u>or again, that</u> M(f) , <u>viewed as a representation of the motivic Galois group, is equivalent to the inflation of a representation of the Taniyama group.</u>

Via I, 5.1, this corollary implies that, for an idempotent e of $K \otimes E_0$ as in 2.2, the motive $e(M(f) \times K)$ is isomorphic to one of the standard motives - see I § 4 - $M(\chi)$ or $M(\overline{\chi})$.

REFERENCES

[AL] G. Anderson, *Logarithmic derivatives of Dirichlet L-functions and the periods of abelian varieties*, Compos. Math. **45** (1982), 315–332.

[A1] G. Anderson, *The motivic interpretation of Jacobi sum Hecke characters*, preprint.

[A2] G. Anderson, *Cyclotomy and a covering of the Taniyama group*, Compos. Math. **57** (1985), 153–217.

[Be] A. Berger, *Sur une sommation de quelques séries*, Nova Acta Regiae Soc. Scient. Upsaliensis **3** (1883), 31 pages.

[Bl] D. Blasius, *On the critical values of Hecke L-series*, Ann. of Math. **124** (1986), 23–63.

[Bl′] D. Blasius, *Period relations and critical values of L-functions*, forthcoming.

[Br] G. Brattström, *Jacobi sum Hecke characters of a totally real abelian field*, Séminaire Théorie de Nombres Bordeaux, exp. 22, année 1981–82.

[BL] G. Brattström and S. Lichtenbaum, *Jacobi sum Hecke characters of imaginary quadratic fields*, Compos. Math. **53** (1984), 277–302.

[BG] J.P. Buhler and B.H. Gross, *Arithmetic on elliptic curves with complex multiplication II*, Inventiones Math. **79** (1985), 11–29.

[CS] S. Chowla and A. Selberg, *On Epstein's Zeta-function (I)*, Proc. Nat. Acad. Sc. USA **35** (1949), 371–374.

[Da] R.M. Damerell, *L-functions of elliptic curves with complex multiplication, I*, Acta Arithm. **17** (1970), 287–301; *II*, Acta Arithm. **19** (1971), 311–317.

[DR] P. Deligne, *Formes modulaires et représentations l-adiques*, Sém. Bourbaki n^o **355** (1968–69).

[DHIII] P. Deligne, *Théorie de Hodge, III*, Publ. Math. IHES **44** (1974), 5–77.

[DP] P. Deligne, *Valeurs de fonctions L et périodes d'intégrales*, Proc. Symp. Pure Math. **33** (1979), part 2; 313–346.

[DB] P. Deligne (texte rédigé par J.L. Brylinski), *Cycles de Hodge absolus et périodes des intégrales des variétés abéliennes*, Soc. Math. France, Mémoire $n^o 2$ (2ème sér.) (1980), 23–33.

[DF] P. Deligne, *Preuve des conjectures de Tate et de Shafarevitch [d'après G. Faltings]*, Sém. Bourbaki n^o **616** (1983–84).

[DMOS] P. Deligne, J. Milne, A. Ogus, and K. Shih, "Hodge Cycles, Motives and Shimura Varieties," Springer Lect. Notes Math., vol. 900, 1982.

[Den] C. Deninger, *On the analogue of the formula of Chowla and Selberg for real quadratic fields*, J. reine angew. Math. **351** (1984), 171–191.

[Ei1] G. Eisenstein, *Beiträge zur Theorie der elliptischen Funktionen: I. Ableitung des biquadratischen Fundamentaltheorems aus der Theorie der Lemniscatenfunctionen, nebst Bemerkungen zu den Multiplications- und Transformationsformeln*, J. reine angew. Math. **30** (1846), 185–210. = Mathem. Werke, I, 299–324.

[Ei2] G. Eisenstein, *Über die Irreductibilität und einige andere Eigenschaften der Gleichung, von welcher die Theilung der ganzen Lemniscate abhängt, — and the sequels to this paper —*, Mathem. Werke, II, 536–619.

[Ei3] G. Eisenstein, *Beweis der allgemeinsten Reciprocitätsgesetze zwischen reellen und complexen Zahlen*, Mathem. Werke, II, 712–721.

[Gi] J. Giraud, *Remarque sur une formule de Shimura-Taniyama*, Inventiones Math. **5** (1968), 231–236.

[GS] C. Goldstein and N. Schappacher, *Séries d'Eisenstein et fonctions L de courbes elliptiques à multiplication complexe*, J. reine angew. Math. **327** (1981), 184–218.

[GS′] C. Goldstein and N. Schappacher, *Conjecture de Deligne et Γ-hypothèse de Lichtenbaum sur les corps quadratiques imaginaires*, CRAS Paris **296** (25 Avril 1983), Sér.I, 615–618.

[Gr1] B.H. Gross, "Arithmetic on elliptic curves with complex multiplication," Springer Lect. Notes Math., vol. 776, 1980.

[Gr2] B.H. Gross, *On the periods of abelian integrals and a formula of Chowla and Selberg*, Inventiones Math. **45** (1978), 193–211.

[Gr3] B.H. Gross, *On the conjecture of Birch and Swinnerton-Dyer for elliptic curves with complex multiplication*, in "Number theory related to Fermat's last theorem," N. Koblitz, ed., Birkhäuser Progr. in Math., vol 26, 1982, pp. 219–236.

[Gr4] B.H. Gross, *Minimal models for elliptic curves with complex multiplication*, Compositio Math. **45** (1982), 155–164.

[Ha] G. Harder, *Eisenstein cohomology of arithmetic groups – The case GL_2*, preprint Bonn 1984.

[HS] G. Harder and N. Schappacher, *Special values of Hecke L-functions and abelian integrals*, in "Arbeitstagung Bonn 1984," Springer Lect. Notes Math., vol. 1111, 1985.

[He] E. Hecke, "Mathematische Werke," Göttingen, 1970.

[Henn] G. Henniart, *Représentations l-adiques abéliennes*, in "Séminaire de Théorie des Nombres, Paris 1980-81," Birkhäuser Progr. in Math., vol. 22, 1982, pp. 107–126.

[Ja] U. Jannsen, *Mixed motives and the conjectures of Hodge and Tate*, preprint.

[K1] N. Katz, *p-adic interpolation of real analytic Eisenstein series*, Ann. Math. **104** (1976), 459–571.

[K2] N. Katz, *p-adic L-functions for CM-fields*, Inventiones Math. **49** (1978), 199–297.

[Kr] L. Kronecker, "Zur Theorie der elliptischen Functionen," Sitzungsber. kgl. preuss., Akad. Wiss. Berlin, 1883, pp. 497–506, 525–530; 1885, 761–784; 1886, 701–780; 1889, 53–63, 123–135, 199–220, 255-275, 309–317; 1890, 99–120, 123–130, 219–241, 307–318, 1025–1029. Werke, vol IV/V, Leipzig-Berlin, 1929/1930.

[Kb] D. Kubert, *Jacobi sums and Hecke characters*, Am. J. of Math. **107** (1985), 253–280.

[KL] D. Kubert and St. Lichtenbaum, *Jacobi sum Hecke characters*, Compos. Math. **48** (1983), 55–87.

[Ku] E.E. Kummer, *Beitrag zur Theorie der Function* $\Gamma(x) = \int_0^\infty e^{-v} v^{x-1} dv$, J. reine angew. Math. **35** (1847), 1–4.

[La] E. Landau, *Über die zu einem algebraischen Zahlkörper gehörige Zetafunction und die Ausdehnung der Tschebyschefschen Primzahlentheorie auf das Problem der Vertheilung der Primideale*, J. reine angew. Math. **125** (1903), 64–188.

[LD] S. Lang, *Relations de distributions et exemples classiques*, Séminaire Delange-Pisot-Poitou, 19e année, 1977/78, n^o **40**.

[LCM] S. Lang, "Complex Multiplication," Springer, Grundlehren, vol. 255, 1983.

[Lg] R.P. Langlands, *Automorphic Representations, Shimura Varieties, and Motives. Ein Märchen*, Proc. Symp. Pure Math. **33** (1979), part 2; 205–246.

[Le] A.M. Legendre, "Exercices de Calcul Intégral," Paris, 1811.

[Ler] M. Lerch, *Sur quelques formules relatives au nombre des classes*, Bull. Sc. Mathém. (2) **21** (1897), prem. partie, 290–304.

[Li] S. Lichtenbaum, *Values of L-functions of Jacobi-sum Hecke characters of abelian fields*, in "Number Theory Related to Fermat's Last Theorem," N. Koblitz, ed., Birkhäuser Progr. in Math., vol. 26, 1982, pp. 207–218.

[Mar] J. Martinet, *Character theory and Artin L-functions*, in "Algebraic Number Fields," A. Fröhlich, ed., Proc. LMS Symp. Durham, Acad. Press, 1977.

[Mom] F. Momose, *On the l-adic representations attached to modular forms*, J. Fac. Sc. Univ. Tokyo **28** (1981), 89–109.

[Mor] C.J. Moreno, *The Chowla-Selberg Formula*, preprint IHES, 1980.

[Od1] T. Oda, "Periods of Hilbert modular surfaces," Birkhäuser Progr. in Math., vol 19, 1982.

[Od2] T. Oda, *Hodge structures of Shimura varieties attached to the unit groups of quaternion algebras*, in "Galois groups and their representations," Adv. Studies in Pure Math., vol. 2, 1983, pp. 15–36.

[Ri1] K. Ribet, *Twists of modular forms and endomorphisms of abelian varieties*, Math. Ann. **253** (1980), 43–62.

[Ri2] K. Ribet, *Endomorphism algebras of abelian varieties attached to newforms of weight 2*, in "Sém. Théorie des Nombres Paris 1979-80," Birkhäuser Progr. in Math., vol 12, 1981, pp. 263–276.

[Ro] D. Rohrlich, *Galois conjugacy of unramified twists of Hecke characters*, Duke Math. J. **47** (1980), 695–703.

[Sa] N. Saavedra Rivano, "Catégories Tannakiennes," Springer Lect. Notes Math., vol. 265, 1975.

[SchA] N. Schappacher, *Zur Existenz einfacher abelscher Varietäten mit komplexer Multiplikation*, J. reine angew. Math. **292** (1977), 186–190.

[SchO] N. Schappacher, *Une classe de courbes elliptiques à multiplication complexe*, in "Séminaire de Théorie des Nombres, Paris 1980/81," Birkhäuser Progr. in Math., vol 22, pp. 273–279.

[Sch1] N. Schappacher, *Propriétés de rationalité de valeurs spéciales de fonctions L attachées aux corps CM*, in "Séminaire de Théorie des Nombres, Paris 1981-82," Birkhäuser Progr. in Math., vol. 38, 1983, pp. 267–282.

[Sch2] N. Schappacher, *Tate's conjecture on the endomorphisms of abelian varieties*, in "Rational Points," by G. Faltings, G. Wüstholz et al., Seminar Bonn/Wuppertal 1983/84, Aspects of Math. (Vieweg) E6, Braunschweig, 1984, pp. 114–153.

[SchΓ] C.-G. Schmidt, *Gauss sums and the classical Γ-function*, Bull. London Math. Soc. **12** (1980), 344-346.

[Schm] C.-G. Schmidt, "Zur Arithmetik abelscher Varietäten mit komplexer Multiplikation," Springer Lect. Notes Math., vol. 1082, 1984.

[dS1] J. de Séguier, "Formes quadratiques et multiplication complexe. Deux formules fondamentales d'après Kronecker," Berlin, 1894.

[dS2] J. de Séguier, *Sur certaines sommes arithmétiques*, J. Mathém. pures et appl. **5** (1899), 55–115.

[SC] A. Selberg and S. Chowla, *On Epstein's Zeta-function*, J. reine angew. Math. **227** (1967), 86-110.

[Sℓ] J.-P. Serre, "Abelian ℓ-adic representations and elliptic curves," Benjamin, 1968.

[ST] J.-P. Serre and J. Tate, *Good reduction of abelian varieties*, Ann. Math. **88** (1968), 492–517.

[SGA4$\frac{1}{2}$] SGA4$\frac{1}{2}$, "Cohomologie étale," par P. Deligne, Springer Lect. Notes Math., vol. 569, 1977.

[ShT] G. Shimura and Y. Taniyama, "Complex Multiplication of Abelian Varieties and its Applications to Number Theory," Publ. Math. Soc. Jap., vol. 6, 1961.

[Shim] G. Shimura, "Introduction to the Arithmetic Theory of Automorphic Functions," Publ. Math. Soc. Japan, vol. 11, 1971. Princeton University Press.

[ShiL] G. Shimura, *On the zeta-function of an abelian variety with complex multiplication*, Ann. Math. **94** (1971), 504–553.

[ShiE] G. Shimura, *On elliptic curves with complex multiplication as factors of the Jacobians of modular function fields*, Nagoya Math. J. **43** (1971), 199–208.

[ShiF] G. Shimura, *On the factors of the jacobian variety of a modular function field*, J. Math. Soc. Japan **25** (1973), 523–544.

[ShiA] G. Shimura, *On some arithmetic properties of modular forms of one and several variables*, Ann. Math. **102** (1975), 491–515.

[ShiM] G. Shimura, *On the periods of modular forms*, Math. Ann. **229** (1977), 211–221.

[ShiP] G. Shimura, *Automorphic forms and the periods of abelian varieties*, J. Math. Soc. Japan **31** (1979), 561–592.

[ShiO] G. Shimura, *The arithmetic of certain zeta functions and automorphic forms on orthogonal groups*, Ann. of Math. **111** (1980), 313–375.

[Sin] W. Sinnott, *On the Stickelberger ideal and the circular units of an abelian field*, Inventiones Math. **62** (1980), 181–234.
[St] M.A. Stern, *Beweis eines Satzes von Legendre*, J. reine angew. Math. **67** (1867), 114–129.
[Tt] J. Tate, *Fourier Analysis in Number Fields and Hecke's Zeta-Functions*, in "Algebraic Number Theory," J.W.S. Cassels and A. Fröhlich, eds., Brighton Proc., London-New York, 1967.
[Wa] W.C. Waterhouse, "Introduction to affine group schemes," Springer GTM, vol. 66, 1979.
[WI-III] A. Weil, "Oeuvres Scientifiques — Collected Papers," three volumes, Springer, 1980.
[WAG] A. Weil, "Adèles and algebraic groups," Birkhäuser Progr. in Math., vol. 23, 1982.
[WBN] A. Weil, "Basic Number Theory," Grundlehren, vol. 144, Springer, 1974.
[WEK] A. Weil, "Elliptic functions according to Eisenstein and Kronecker," Springer, 1976.
[WW] E.T. Whittaker and G.N. Watson, "A Course of Modern Analysis," Cambridge, 1952.

ALPHABETICAL LIST OF SYMBOLS AND CONCEPTS

The following is mainly a list of notation in alphabetical order, according to the usual english transliteration of the symbols. - Are also listed important concepts used in the text which do not have an established or easily guessed symbol attached to them, and which do not occur in one of the headings of the table of contents.

A.

A	abelian variety (or elliptic curve) with complex multiplication (CM)	I,1;III,1
a_D	element of $\mathbb{B}^0_{\mathbb{Q}(\sqrt{-D})}$ such that $J_D = J_{\mathbb{Q}(\sqrt{-D})}(a_D)$	0,8.3
	affine group scheme	I,2.3.1
$\mathcal{AH}$	category of arithmetic Hodge structures	I,7.3.1
$A_k, \overline{A}_k$	Kuga-Sato variety, compactified	V,1.1.1
α	adelic splitting of Taniyama group	I,6.5
	Anderson's theory of Jacobi sums	0,8.2-4;I,7; II,4.1-3
$\tilde{A}_{\mathfrak{p}}$	reduction of A at $\mathfrak{p}$	I,1.3
$\mathrm{Aut}^{\otimes}\omega$	tensor automorphisms of fibre functor	I,2.3

B.

$\mathbb{B}, \mathbb{B}^0, \mathbb{B}_K, \mathbb{B}^0_K, \mathbb{B}_{(p)}$	free abelian group on $\mathbb{Q}/\mathbb{Z}\setminus\{0\}$, and certain subgroups thereof	0,8.2.1-2
	Brauer's induction lemma (generalized, applied to $\mathfrak{T}$)	I,6.5.8

C.

$c^{p}_{AH}(X/K)$	absolute Hodge cycles of codim p on X, defined over K	I,2.1
χ	Hecke character (of K with values in E)	0,1
$\chi_{\mathbb{A}}$	idèle character associated to χ	0,5
$\chi_{D,\mathfrak{p}}$	D-th power residue symbol modulo $\mathfrak{p}$	0,8.1
χ_{λ}	λ-adic idèle class character of χ	0,5
	1-dim. λ-adic Galois representation	I,1.3-5
χ^{τ}	Größencharaktere associated to χ	0,6
	"Chowla-Selberg-formula" (= Lerch's formula)	III,1.4.4; III,2;III,3.3
$\mathcal{CM}_K$	category of motives over K generated by potentially CM abelian varieties	I,4
$\widetilde{\mathcal{CM}}$	category of Anderson's ulterior motives	I,7.3.3
	corrigenda	II,1.7.11; II,3.1.1 II,3.4
$c^{\pm}(\chi)$	periods of χ	II,1.8
$c^{\pm}(M)$	Deligne's periods of the motive M	II,1.6
	critical integer	II,2.0
$\mathrm{Crit}_K(\underline{a})$	(half of the) critical integers for $J_K(\underline{a})$	II,4.2
$\mathcal{CV}_K$	category of the h(X)'s, X a variety over K	I,2.2
	cyclotomic character: see Ψ	

D.

D	-disriminant of $\mathbb{Q}(\sqrt{-D})$	0,8.1;III,1
$\mathfrak{d}$	discriminant of extension of number fields	II,1.4.8

$D^{\pm}$	fudge factor in comparison of p and $c^{\pm}$	II,1.7.3-12
	Deligne's rationality conjecture	II,2.1; V, Introd.
$\delta(K'/K,\sigma)$	fudge factor for p(M) under restriction of scalars	II,1.4.2-8
Δ_{μ}	fudge factor in refined "Chowla-Selberg-formula"	III,1.3.2; III,1.4.4-5; III,1.5.6; III,3.2
$\delta^{\pm}$	fudge factor in comparison of p and $c^{\pm}$	II,1.7.3-12
$\det_E(M)$	determinant motive	II,1.1.2
D_{σ}	fudge factor of p(M) under restriction of scalars	II,1.4.2-8
<>	representative in $\mathbb{Q}$ of $\mathbb{Q}/\mathbb{Z}$	0,8.1.4

E.

E,E',E_0	fields of values/coefficients (always of finite degree over $\mathbb{Q}$)	0;I;II;V
$\mathbb{E},\mathbb{E}_m,E_m$	"Euler's" arithmetic Hodge structures	I,7.3.2; I,7.4.2; I,7.4.5
$\mathrm{End}_K A$	endomorphisms defined over K of A	I,1.1
ϵ	- adelic splitting of S_K	0,7.4
	- Dirichlet character of $\mathbb{Q}(\sqrt{-D})$ - character of theta series f	0,8.1;III,1 V,2.0

F.

f	newform on $\Gamma_0(N)$ of weight > 1	V
$\mathfrak{f}$	conductor or defining ideal	0,1
f_E	Tate's class field theoretic "cocycle"	I,6.4.2;I.6.5
Frob $\mathfrak{p}$	geometric Frobenius element	I,1.2

G.

$\Gamma,\Gamma(\underline{a})$	gamma function/product of special values thereof	II,4.0; II,4.2.3
g_E	Tate's "cocycle" derived from reciprocity law for CM motives	I,6.3.1; I,6.5
$g_K(\underline{a},\mathfrak{p})$	Jacobi sum	0,8.2.3
g_p	Gauss sum map	0,8.2.2
$G(\sigma)$	motivic Galois group (relative to H_σ)	I,2.3.4

H.

H	Hilbert class field of $\mathbb{Q}(\sqrt{-D})=K$	III,1
	H/K-curves	III,1.5.2
$H_A, H_{A^f}, H_{DR}, H_\ell, H_\sigma$	various cohomology theories	I,2.1
h_D	class number of $\mathbb{Q}(\sqrt{-D})$	0,8.1;0,8.4.3; III,1.0
	height function (Faltings')	III,2
$\mathcal{H}od_{\mathbb{Q}}$	category of rational Hodge structures	I,2.4.2; I,6.0-1
h(X)	the motive of a variety	I,2.2

I.

I, I_σ	comparison isomorphism	I,2.1.1; I,7.3.1; II,1.1.1; II,1.2.3; II,1.6.3
	invariance lemma of periods under extension of the field of definition	II,3.2.7

J.

J_D	basic Jacobi sum Hecke character of $\mathbb{Q}(\sqrt{-D})$	0,8.1;0,8.3
$J_K(\underline{a})$	general Jacobi sum Hecke character	0,8.2.4

K.

K	field of definition (of finite degree over $\mathbb{Q}$)	passim
	Karoubian envelope	I,2.2
	$\overline{K}/K$-forms of rank-1 motives	II,3.5
	Kummer's series for $\log \Gamma$	III,2

L.

$L(A/K,s)$	(false) Hasse-Weil L-function of an abelian variety A over K	I,1.7
$\Lambda(\chi^\tau,s)$	Hecke L-function with Γ-factors	0,6
$L(\chi^\tau,s)$	Hecke L-function	0,6
	Legendre's conjecture	III,3.1.1
	Legendre's period relation generalized	II,1.5.4
	lemniscatic arc length	I,7.5;V,2.4.1
$L^*(A/K,s)$	$E\otimes\mathbb{C}$-valued L-function of A with complex multiplication	I,1.7
$L^*(\chi,s)$	$E\otimes\mathbb{C}$-valued array of Hecke L-functions	0,6
$L^*(M/K,s)$	$E\otimes\mathbb{C}$-valued L-function of a motive in $\mathcal{CM}_K(E)$	I,6.5.9
$L_{usual}(A/K,s)$	Hasse-Weil L-function of an abelian variety A over K	I,1.7

M.

$M(\underline{a})$	motive for special Jacobi sum Hecke character	I,7.1
$M(\chi)$	(standard) motive for the Hecke character χ	I,4-5
$M(f)$	motive for newform f	V,1.1
$\mathcal{M}_K$	category of motives over K	I,2.2
$M_K(\underline{a})$	motive for Jacobi sum Hecke character	I,7.2
$\mathcal{M}_K^{av}$	category of motives obtained from abelian varieties	I,2.4.2
$\dot{\mathcal{M}}_K$	false category of motives over K	I,2.2
$\dot{\mathcal{M}}_K^+$	false category of effective motives over K	I,2.2
$\mathcal{M}_K(E)$	category of motives over K with coefficients in E	I,3
$\mathcal{M}_K^0$	category of Artin motives over K	I,2.4.1
	modular proof of Deligne's rationality conjecture (hypothetical)	V,Introd.
	monomial period relations (Shimura)	II,4.4.3;IV,1
	motivic Galois group: see $G(\sigma)$	
MT(A),MT(V)	Mumford-Tate group of an abelian variety/a rational Hodge structure	I,2.3.3; I,6.0.2
μ	- cocharacter of a rational Hodge structure - Hecke character of finite order (= "Dirichlet character")	0,7.3.4; I,6.0-1 II,3;III,1; IV,2

N.

$n(\sigma,\tau)$	Hodge coefficients of a CM-structure	0,4;I,1.7; I,4.2;I,5.1; I,6.1.5
n_1,n_c	coefficients of T_D	0,8.1.3;III,1
	number of CM-type	0,3

P.

$p(\chi)$	period of the Hecke character χ	II,1.8
	period conjecture of Gross	IV,2.1.7-9
$p(M)$	period of the motive M	II,1.1
	pseudoabelian envelope	I,2.2
Ψ	the cyclotomic character	0,8.2.1

Q.

	$\mathbb{Q}$-curves	I,7.5; III,1.5.4
$\mathbb{Q}(\pm 1),\mathbb{Q}(n)$, $\mathbb{Q}_{DR}(n)$	powers of Tate motive, and its realizations	I,2.1
	quotients of L-values generating abelian extensions	II,3.3.0

R.

	"reflex motive" of Blasius	II,2.2
	reflex principles	IV,1.2.11
$\mathcal{R}ep_K(G)$	category of finite dimensional representations of an affine group scheme G	I,2.3
	"root numbers" (of Dirichlet characters) as twisting factors	II,3.4

S.

$\mathbb{S}$	$= R_{\mathbb{C}/\mathbb{R}}\mathbb{G}_m$	I,6.0
	semi-simple (tannakian) category	I,2.2
	the Serre group: see Z, Z_K	
$S_K = \varprojlim_f S_{K,f}$	Serre's group	0,7.2-4

St_K	Stickelberger ideal	0,8.4
	strong Weil curve for $\Gamma_0(27)$	I,7.5
	strong Weil curve for $\Gamma_0(32)$	I,7.5;V,2.4.1

T.

T	- infinity type - CM-type of abelian variety	0,1-3 I,1.6
$\mathfrak{T}, {}_K\mathfrak{T}$, $\mathfrak{T}_E, {}_K\mathfrak{T}_E$	Taniyama group, and subquotients thereof	I,6.4-5
$\tilde{\mathfrak{T}}$	group scheme of $\widetilde{\mathcal{CM}}$	I,7.3.3
T	Hecke algebra	V,1.1.3
	tannakian category (neutralized)	I,2.2;I,2.3.1
	Tate twist/Tate motive: see $\mathbb{Q}(n)$	
T_D	infinity type of J_D	0,8.1.5; 0,8.3;III,1
	tensor category	I,2.2
θ_K	map used in writing Stickelberger elements	0,8.2.1
$T_\ell(A)$	Tate module of A	I,1.1
T_p	Hecke operator	V,1
	transcendence theory	I,1.4;II,4.0; III,3.1.1

U.

$\mathfrak{U}$	group scheme for $(\mathcal{CM}_{\mathbb{Q}}, H_B)$	I,6.3;I,6.5

V.

V_E	generalized "half-transfer" of Tate	I,6.4.0; II,3.2.8
$V_\ell(A)$	$= T_\ell(A) \otimes_{\mathbb{Z}_\ell} \mathbb{Q}_\ell$	I,1.1

W.

W	arithmetic Hodge structure	I,7.3.1
$w, \tilde{w}$	weight	0,3;0,7.3.4; I,6.0-1

X.

X_m^n	Fermat hypersurface	I,7.1;II,4.1
$X_1(N)$	closed modular curve	V,1.1.1

Y.

$Y_1(N)$	open modular curve	V,1.1.1

Z.

$Z=\varprojlim Z_K$	the Serre group	0,7;I,6.1.4
$Z_{K,f}$	finite level of the Serre group	0,7.1

MIX
Papier aus verantwortungsvollen Quellen
Paper from responsible sources
FSC® C105338

If you have any concerns about our products,
you can contact us on
ProductSafety@springernature.com

In case Publisher is established outside the EU,
the EU authorized representative is:
Springer Nature Customer Service Center GmbH
Europaplatz 3, 69115 Heidelberg, Germany

Printed by Libri Plureos GmbH
in Hamburg, Germany